REPATRIATION, INTEGRATION or RESETTLEMENT

REPATRIATION, INTEGRATION

or

RESETTLEMENT

The Dilemmas of Migration among Eritrean Refugees in Eastern Sudan

Sadia Hassanen

The Red Sea Press, Inc.

Publishers & Distributors of Third World Books

P.O. Box 1892
Trenton, NJ 08607

P.O. Box 48
Asmara, ERITREA

The Red Sea Press, Inc.
Publishers & Distributors of Third World Books

RSP

P.O. Box 1892
Trenton, NJ 08607

P.O. Box 48
Asmara, ERITREA

CIP data is available from Library of Congress

ISBNs:
1-56902-273-9 (hard cover)
1-56902-274-7 (Paperback)

CONTENTS

ACKNOWLEDGMENTS

I thank God for giving me all that I have in my life and for enabling me to finish this book. Writing this book has not been an easy process. During the years I have been struggling, there were many people and organisations that directly and indirectly supported me and helped me to finish it. I would like to thank all those who worked hard and side-by-side with me as well as those who helped me to think and consider other things in life. This study is about Eritrean refugees in the town of Kassala in eastern Sudan. Thus, without the participation of the refugees in Kassala, this study would not have taken place. I thank them for making this study happen. Beside the refugees, other institutions and actors also helped in conducting this study in Sudan. I thank all those who supported me and hosted me in Sudan. Special thanks go to COR Khartoum, Kassala, and Shawak. Ibrahim Abdullah, Hader Yousif and Mawalan. I might have forgotten some of names but this does not mean that their help is less credible than the mentioned ones. I would like to thank the Department for Research Cooperation (SAREC) for financing this research, and the Centre for international migration and ethnic relation (CEIFO) as well as the department of Human geography in Stockholm. Without the support of these institutions, this study would have been impossible. Many people from both departments were involved in this study as well as friends and work colleagues. I thank them all. However, some of them deserve special attention. My very special thanks go to my supervisors Professor Charles Westin and Associate professor Gunilla Andrae, Professor

Bo Malmberg, Dr. Shahram Khosravi, Dr. Lars Wåhlin and Lotta Wistedt. Other people outside these two departments were also involved on the work of this project, and my special thanks go to Associate professor Irena Molina, Dr. Åsa Gustafson, Professor Lars Dahlgren, Dr. Hauwa Mahdi, and Dr. Lowe Börjesson. Idris Hassan, Ibrahim Gadam, Mahmued Adem Ibrahm, Dr. Redie Bereketab. From the Eritrean home pages, I thank the Awate home page, although the page provides much vital information about Eritrea's political, social and economic conditions. My gratefulness to it is not about these issues but it is about Salh Gadi's funny and valuable stories. Thank-you "kreneno". In the private sphere, I would like to thank my family, who always supported me in my life despite the geographical distance that came between us. My first special thanks go to the Afa family, starting from my mother until Abdulkarim and then starting from Nuha up to Sumaya. Thank you. Without you, life would have been sour. My last gratitude goes to the friends and families in Uppsala, Stockholm, Umeå, London, Riyadh and Melbourne.

Sadia Hassanen

This Book is dedicated to

Kedeja Hammed
and to the memory of
Hassanen Afa (1941–1984),
Haja Drar and Amna Osman

ABBREVIATIONS

CERA	Commission of Eritrean Refugee Affairs
COR	Commissioner For Refugees
EIJM	Eritrean Islamic Jihad Movement
ELF	Eritrean Liberation Front
EPLF	Eritrean People's Liberation Front
ERREC	Eritrean Relief and Refugee Commission
NDA	National Democratic Alliance
NIF	National Islamic Front
PDFDJ	People's Front for Democracy and Justice
REST	Relief Society of Tigray
RSD	Refugee Status Determination
SOAS	School of Oriental and African Studies
SPLA	Sudan People's Liberation Army
SPLM	Sudan People's Liberation Movement
TPLF	Tigray People's Liberation Front
UNDP	United Nations Development Programme
UNHCR	United Nations Higher Commissioner for Refugees

GLOSSARY

Abukamsa:	a zone nearby Muraba'at where there is heavy concentration of Eritreans (one of the areas of the study)
Beja:	coalition of clans in eastern Sudan
Beni Amer:	clans that are settled in Eritrea and Sudan
Butaka:	identity card (in Arabic)
Gedaref:	one of the main towns in Eastern Sudan, known for its agri- cultural production, mainly durah (sorghum) and sesame.
Halfa:	a small town outside of Kassala town where there is an irrigation scheme growing cotton.
Jinsia:	nationality (in Arabic)
Kabasa:	highlanders, Tigrinya speakers
Legina:	refugee committee in the refugee camps
Muraba'at:	a zone in Kassala town mostly inhabited by people from Eritrea (one of the study areas).
New caseload:	refugees who left Eritrea after independence

Old caseload:	refugees that left Eritrea before its independence
Omeda:	head of clan (in Arabic)
Omedia:	the lowest unit in the structure of local administration based on common tribal affiliation
Refugee camps:	transit centres where refugees are kept until a durable solution is found in the first country of asylum
Shari'a:	Islamic law
Suk-Kassala:	Kassala market (one of the study areas in Kassala)
Tigre Speakers:	all clans that use Tigre as their first language
Umma:	the worldwide Moslem community or nation

PART 1

BACKGROUND INFORMATION
Theory and Methodology

This thesis has eight chapters. Chapter one introduces the study subject and the UNHCR policies. Chap ter two defines the main concepts and terminologies and presents the conceptual framework. Chapter three presents the methodology of the study and the diverse sources on which the thesis is based. It also discusses in detail the research process, the problems encountered and the approaches used to overcome them.

Chapter four is divided into two sections. The first section provides some background data on Eritrea that show the extent to which Eritrean society is diverse in terms of ethnicity, religion, language and political orientations. It also presents a brief description of gender relations and the historical links that exist between the border communities of the two countries of Eritrea and Sudan. The second section presents some background data on Sudan and Kassala town, the refugees' country of asylum and place of residence. In view of the fact that religion and ethnicity are some of the factors that influence local integration in Sudan, the chapter discusses the issues of Islam and ethnicity.

Chapters five to seven, Part Two of the book, present the findings of the study. These three chapters identify the political, social and economic factors that influence the decisions of the refugees to return to their country of origin, to remain in the country of asylum or to migrate to third countries according to my survey. Chapter eight presents the conclusions of the study and draws some lessons that may have policy implications.

1

INTRODUCTION

This study is about Eritrean people who left Eritrea to seek asylum in Sudan and over time settle there as refugees. The Eritrean war of independence forced thousands of Eritreans into exile. The war of independence ended in 1991. Since then some Eritrean refugees have voluntarily returned while others have preferred to stay in Sudan. This study is about those who decided not to return. Before I present the aim and the scope of the study, I will introduce some background information about refugee conditions in Africa and UNHCR policies.

In recent decades, millions of people in Africa have been forced to leave their countries of origin as a result of political, social and economic disasters. Africa is a continent where a large share of today's refugees and displaced persons are located. Of the twenty largest refugee-producing countries, nine are located in Africa (Crisp, 2000).

In the 1960s and 1970s, African countries stood for a liberal policy towards the refugees; this was the time when

African refugee reception was labelled the "Golden age" of asylum in Africa (Crisp, 2000, p. 4). For various reasons, the refugee situation in Africa is very different today. Many African countries are reluctant to receive refugees and the reason for this is their poor economy and the absence of democratic government.[1] The refugees from the 1960s and 1970s had other motives to migrate that were linked to the struggle against European colonialism in Africa. This was the time when Pan-Africanism and anti-colonialism was strong throughout Africa, and influential political leaders such as Julius Nyerere and Kenneth Kaunda strongly supported the idea of liberal refugee policies. Moreover, the economy of many African states at that time was not as entirely devastated as it is at present. Host countries had the ability to handle the economic burden imposed on them by the presence of refugees from neighbouring and nearby states (Crisp, 2000, p. 4).

Unlike in the past, practically all-current forced migration is caused by internal conflicts and instabilities. In recent years conflicts that occur within countries, e.g. in the Horn of Africa (Ethiopia, Eritrea, Djibouti, Somalia and Sudan) have received much attention. The cost of these conflicts has been high for the countries involved. The increasing costs of war and the loss of productivity have been ruining the economies and societies of the Horn of Africa. Wars and conflicts in the Horn of Africa have been some of the major causes of poverty and population displacement. Eritrea is one of the countries that have been hard hit by the damaging consequences of war. Refugee hosting African countries are economically poor.

By referring to the poor economy of the country as a justification, many African countries have adopted a policy of placing refugees in some peripheral region, mostly in rural areas. On top of that, refugees in these countries do not have secure situations. These countries do not have any clear policy concerning a solution for the refugees, for instance refugees do not have the legal right to be settled at any place they want to in the receiving country. By contrast, refugees who are settled in the countries of the north and in Australia have the right to be integrated, and after being settled for some time, they have the right to become citizens of those countries (Crisp, 1999, 2000). The United Nations High Commissioner for refugees (UNHCR) provides economic aid and health care to refugees who are settled in the countries that have poor economy.

The refugees I have met during the course of this study are not supported by any agency or authority in the host country. These are refugees that well illustrate the current dominant refugee situation in Africa, namely that refugees are people who have left their country as the result of conflict within their country and who are settled in a neighbouring country with a poor economy. My definition of the refugee concept is broad. It is psychological in the sense that it refers to people's self-identification as refugees. It is not a legal definition; it does not follow the definition used by the UNHCR. Regardless of how "refugees" define themselves or how long they reside in the host countries, these people live in a state of limbo precisely because most African host countries lack naturalisation policies. This certainly applies to the Eritrean refugees living in Sudan. Thus, I use the notion refugee/Eritrean refugee to refer to all Eritreans who identify themselves as people who due to human rights violations have fled from their country of origin. This is despite the fact that they do not receive any kind of help or and despite the fact that they are not recognised as being displaced as those in the refugee camps are.

The International Community and its Policies

The poor economy of most African countries makes African refugees dependent on the United Nations High Commissioner for Refugees (UNHCR) for financial support as well as for future solutions to their situation. The UNHCR was established to protect refugees and to provide them with humanitarian assistance. The agency was initially mandated to protect refugees and displaced persons in the chaotic aftermath of World War II and who resided in Europe. The refugees were expected to be resettled permanently in North America and Western Europe, and they were expected by host governments to remain and take their place within the new society (Crisp, 2002). Today's refugees are mostly from countries of the south, that is to say from economically poor countries, and they seek asylum in other economically poor countries. As a consequence of this situation donor governments and international humanitarian agencies shape the policies of host governments in the south. Moreover, as their numbers are growing it has become important for the international community to find solutions to the problem of forced migration (ibid.).

According to Moussa (1993) the label refugee never changes until those who have fled return to their countries of origin. According to this author the label refugee is central to the formation of social and economic policies of host governments. She also notes that the word refugee problem represents a relationship of inequality between givers (NGOs, governments, the United Nations etc) and recipients (refugees). Moussa further argues that when people are reduced to dependence on aid givers, their sense of agency is circumscribed (ibid.).

Permanent solutions

The office of the UNHCR, whose mandate includes the responsibility of finding permanent solutions to the refugee problem, has been promoting three strategies known as local integration, resettlement to a third country and voluntary repatriation. These are regarded as durable solutions. However, African governments, for whom the concept has been suggested, have not welcomed idea of local integration into the host country (Crisp, 2004).

According to the UNHCR, repatriation should take place voluntarily and in conditions of safety and dignity. Return should only take place if fundamental changes in the circumstances causing displacement have come about. For instance, a peace agreement may have been signed, or a violent regime replaced by a new government that protects the rights of its nationals (Bakewell, 1999, p. 4; Black and Koser, 1999).

The international community, with the UNHCR in the forefront, regards voluntary repatriation as representing the best solution to the refugee problem. This is underpinned by the assumption that refugees want to return home subsequent to the elimination of the factors that forced them to flee. The discourse on migration/displacement is then about the discourse of place. The corollary of this is that an individual is linked to a particular place or community within a nation state. In addition, the issue of repatriation is based on the assumptions that each state is responsible for its people (Hammond, 1999). Solutions to the refugee problem are therefore designed to enable the refugees concerned to return to their particular places and communities of origin in order to assume their identities.

Its Statute and the 1951 UN Convention Relating to the Status of Refugees govern the UNHCR's activities, in countries of both asylum and return. Additionally, the agency works within the limits set by the countries of origin and asylum. Tripartite agreements signed among the three parties, the UNHCR, governments of countries of asylum and those of origin govern most repatriation operations. The rights and duties of refugees, as well as of the UNHCR, host governments and governments in countries of origin are spelled out in detail in such agreements. The cooperation of governments in countries of origin and asylum is therefore critical to the implementation of successful repatriation operations as well as of reintegration programs.

The role of the UNHCR policies regarding return migration is significant to this study, because one of the research questions is to examine how durable a solution repatriation is.

A Migrant Researcher

How has the migration process affected people's identity and image of what "home" is? Using my own experience and the research findings of other researchers, I shall explore the meaning of home. It is hoped that this will enrich our understanding of the position some of the Kassala refugees take regarding the UNHCR policy.

My decision to engage in this study is partly motivated by my own background and experience as a refugee. The reasons that forced me to flee from my country are almost identical to those which forced my informants to become refugees in Sudan. Being once a refugee myself, the study rekindles many unhappy but also happy memories that I experienced during the long journey I made to reach my present home. To reach this home I had to travel spatially, socially and metaphorically. In researching this project, my memories of many families and friends with whom I grew up, and later shared a brief life in Sudan with,are renewed. I ask myself ever so often, how and why my life turned out to be what it has become and how my identities have been affected over time by migration and the multiple experiences associated with it.

Before engaging in this research project, home for me was a fixed a place. The idea of undertaking research on "return migration" was born in 1999 when I was working on a master's dissertation in human geography. The aim of the master's thesis was to examine the impact of migration on gender relations among Blin[2] immigrants in Melbourne, Australia. In the process of interviewing different respondents, it became clear that my interviewees had different perceptions of "home". When asked to explain what home meant to them some said: "My place is where my children are." Others said: "Home is where I can use my real identity as an Eritrean. Here in Australia, I live and work as an Eritrean. In Sudan, I was not allowed to live or work as an Eritrean. I used Sudanese identity in order to gain access to work."

These responses were eye-openers to the some of the factors that affect our perception of home. They enabled me to question my own understanding of the meaning of home and other things, which I previously had taken for granted. These encounters made me reflect on my own life. You can imagine the impact of those responses, having lived in Sweden without ever visiting Eritrea or even thinking about the changes I might have undergone. Prior to this research, I took my identity for granted. In Sweden, the question of who you are and where you come from is part of everyday life as an immigrant. One is constantly reminded of one's 'otherness.' Every time people ask me about my origin, I proudly, say that I am an Eritrean and that I came to Sweden as a refugee. When confronted with the question: "When will you return to your country," I used to say: "When I complete my studies." Although I routinely said so, I never had any concrete plan regarding when to return or where exactly to return to. However, every time I was asked such a question, I repeated the same cliché. Questions such as: "When will you return to your country?", no matter how benign, serve as reminders that I am in a place that does not belong to me and reinforces my 'otherness.' It is as if the people who ask such questions do not approve of my belonging to Swedish society. They probably think that since I do not originate from Sweden and do not share the visual physical attributes of an ethnic Swede, it is natural for me to feel that I do not to belong here and therefore want to return to where I belong.

The findings of the study in Melbourne made me rethink the meaning and significance of 'home,' 'belonging,' and 'return migration' among refugees. It was such reflections that motivated me to work on these issues in greater depth and hence the idea of writing a PhD thesis on the refugees living in Kassala, my first refugee "home". It is with these theoretical issues surrounding migrants and the questions of their identities that I approach what other researchers have done about the specific case of Eritrean refugees.

The Scope of the Study and Departure Points

This study follows the case of Eritrean refugees in Sudan. This country has hosted Eritrean refugees since 1967 (Kibreab, 1990; Ibrahim, 2003; Kardawi, 1999). The particular groups of refugees in this study are the Eritrean refugees who live in the town of Kassala. It is important to point out that the results of this study only address refugees who live in urban settings in eastern Sudan, specifically those who are settled in the town of Kassala. This is important to point out because the situation of the refugees in urban areas differs significantly from conditions for those living in refugee camps.

One point of departure for the thesis is to look into what the effect of political change in Eritrea may have had on people's willingness to return. Does the elimination of the factors that caused the initial flight have anything to do with the decision-making concerning return migration? Moreover, the study questions the assumption that home is linked to a specific location and that people are linked to that location naturally. Furthermore, this study questions the assumption that implies that "once a forced migrant, always a forced migrant" and that return migration is the only durable solution to refugee problems.

Another point of departure is the work of Gaim Kibreab, an associate professor in the department of development studies at the University of South Bank, London. He is the author of several books and articles on refugee issues in Africa, especially in Sudan. Since thirty years he has studied the situation of Eritrean refugees. In 1996, he conducted a study in Kassala on voluntary repatriation of Eritrean refugees from Sudan. The study investigated the factors influencing the refugees' decisions to return or to remain in Sudan. The results of his research were published in his book with the title "Ready and willing, but still waiting". Kibreab's study took place six years before my own study was conducted. Kibreab shows that the decision of the majority of Eritrean refugees concerning repatriation subsequent to the elimination of the factors that prompted them to flee was a function of interplay between economic and environmental factors. He argues that since the large majority of the refugees lived in poverty, the question of whether they stayed in Sudan or returned to Eritrea was to a large extent determined by livelihood considerations upon return. His study also shows that there were some, but a clear minority, whose decision to stay in Sudan was motivated by political considerations. The majority of the latter were former ELF combatants and members of the ELF. So what has happened since Kibreab's study? Are the Eritrean refugees in the town of Kassala still ready and willing to return as they were then? (Kibreab, 1996).

Aim and Research Questions

This study focuses on the situation of urban Eritrean refugees in Kassala town in eastern Sudan. The central aim of the thesis is to identify and analyze the factors that influence the refugees' decision concerning return migration.

The research questions this study seeks to address include: Why have the Eritrean refugees remained in Sudan instead of returning to Eritrea despite the political changes that occurred there in May 1991? What are the major factors that influence their decision? To what extent do the political changes that occurred in Eritrea allow the refugees to return in safety and dignity? How do the refugees look at the political changes that occurred in their country of origin? Most of the Eritrean refugees came from rural settings in Eritrea but settled in urban areas in Sudan. What has been the impact of this displacement on gender relations among the refugees? To what extent is the decision of the refugees not to return to Eritrea influenced by the socio-economic conditions in Sudan? How do the changes that refugees have gone through in exile affect their decision to return to Eritrea, to remain in Sudan or to emigrate further a field?

In 1991 the regime that had forced many Eritreans to leave their country was overthrown by the combined efforts of the liberation fronts after the long war of liberation. A referendum was held in 1993, after which Eritrea became an independent country and a new regime came into power. Despite this change and the opportunity to return, many Eritrean refugees who had left the country before independence chose to remain in Sudan. The question that this thesis addresses is why did these refugees remain in Sudan? In fact, despite of the political change in 1993, Eritrea remains one of the principal refugee producing countries in the Horn of Africa region. In other words, independence as a political achievement of the country neither stopped the flow of new refugees, nor has it led to the return of all those who were displaced before and during the long war of independence. The question that presents itself is: What does this imply?

Studies conducted in Sudan show that the government of Sudan does not want Eritrean refugees to live in towns and yet it does nothing to prevent Eritrean immigrants from settling in Kassala and working in the town. The study aims to understand how the Eritrean settlers cope with this uncertainty.

The whole issue of forced migration and solutions to the refugee problem in terms of settlement or repatriation is highly political. What dimensions of the political sphere are concerned? After all, the economic and social questions that other researchers have raised are to some extent also political issues. To narrow down the study to a specific dimension of the political sphere, the questions will centre on the Kassala refugees' conditions regarding return migration. So the questions are: Can the political reasons for the failure of return migration be that the new government of Eritrea does not protect the democratic and human rights of its nationals? Alternatively, is it because the refugees in Sudan prefer the political system of their host country? Could it be that they have acquired new ideologies, for instance new political ideologies? Whatever the reasons are, these questions and the continued migration of Eritreans suggest that the process of deciding whether to return to one's country of origin or not are complex.

Several previous studies argue that the reason why refugees are dependent on international community and host countries solutions is their helplessness and weakness. The question that this thesis addresses is: Are the refugees in Kassala dependent and weak?

As pointed out above, Gaim Kibreab conducted a study in the town of Kassala in 1996. What comes across from his study is that most of the refugees consider the matter of return migration primarily from an economic point of view. His approach seems to have been to prove the significance of the economic factor. In fact, it might have been a very important factor in the positions the refugees took. One must wonder what economic advantages Kassala had that were so significantly beneficial to them and that they expected that they could not attain in Eritrea. If in fact their economic standards were so much better in Sudan than what they could hope to recreate in Eritrea, why was that the case? What was the political background of the informants?

These are some questions that Kibreab could have taken into account in his analysis of the responses he got from the refugees. Not having addressed these questions seems to have contributed to Kibreab's underestimation of the impact of the problematic relations between the ELF (The Eritrean Liberation Front) and the EPLF (The Eritrean People's Liberation Front) as a factor in the position of refugees on return migration.

While economic considerations remain important to the decisions the refugees make, this study intends to incorporate the political calculations that form part of their strategy. It stresses the role of political affiliation on the decision making regarding return migration.

NOTES:

1. Interview with one of the UNHCR staff in Sudan, 2002.
2. One of the ethnic minority groups in Eritrea.

2

STUDIES IN THE FIELD OF MIGRATION

INTERNATIONAL MIGRATION

In this chapter I shall present previous studies which provide concepts that will help us understand the field of migration.

The field of research known as international migration covers all types of mobility that take place across politically recognized national borders. Hence, if somebody crosses a national border he or she can be classified as an international migrant. Numerous studies within the social sciences have been carried out in order to explain why migration takes place. Most of these studies focus on voluntary types of migration (Gilbert and Gugler, 1982; Westin, 1999). More recently studies about migration deal with both the voluntary and forced migration (Bakewell, 1999, p. 6).

Today, migration is viewed as a universal phenomenon that affects all societies. Earlier studies on involuntary migration mostly dealt with the dynamics of refugee flights and the challenges that refugees faced in connection with the processes of displacement and resettlement in receiving countries. New concepts and perspectives are emerging with recent studies of migration. Within the discourse of migration, the perception of location and the notions of sedentariness and mobility have received greater attention than before (Malkki, 1990; Stepputat, 1994; Warner, 1994).

Migration is a dynamic process and from time to time it takes on different forms. Thus, both voluntary and involuntary forms of migration are complex and dynamic. As some researchers have pointed out, the dynamic nature of the phenomenon necessitates the adoption of non-static approaches that recognize the particularities of each given migrant situation (Westin, 1996). In the area of involuntary migration, conflicts and human rights abuses stemming from bad governance have become the main reasons that produce refugees in many of sub-Saharan African countries (Westin, 1998.).

The concepts of voluntary and involuntary migration refer to different types (and causes) of mobility. The distinction between voluntary migration and forced migration is based on whether physical threats to the person are involved or not. In the case of voluntary migration, the person is not under threat, or subjected to political persecution. This means that the voluntary migrant is not forced to abandon her/his place of origin by life-threatening circumstances. In voluntary migration, the economic interest of the migrant is the driving force behind mobility. In involuntary migration, the economic motives of the migrants are reduced by life-threatening circumstances and therefore people flee in search of safety and security (Moussa, 1993; Sorensen, 1990).

The differences drawn between voluntary and involuntary migrants seem to rest on giving economic reasons for the former's migrant status and political reasons for the latter. Some researchers have questioned the usefulness of differentiating between political (involuntary) migration and economic (voluntary) migration (Malmberg, 1997; Koser, 1997; Sorensen, Van Hear and Pedersen, 2002).

Critics have argued that making this clear cut distinction between voluntary and involuntary forms of migration lacks firm basis and that in some cases the difference is even non-existent. A certain issue that was regarded as political might under changed circumstances become economic and vice versa. The division of migrants into these two groups is for me one of the important issues in the debate because they bear on this research in at least two ways. Firstly, being considered either voluntary or involuntary migrant will decide whether the policy of return migration applies or not. Secondly, the debates on the two categories of migrants also bear on issues of identity and belonging, which are vital to the way we perceive their condition in host countries, as well as in their country of origin (Westin, 1999). To migrate in order to search for better job opportunities will usually involve some comparison of conditions in the places of origin and destination. However, labour migration may vary considerably between different societies. In developed countries, labour migration has had an inter-urban character, while in the developing countries most labour migration takes place from rural to urban areas (Malmberg, 1997). Labour migration is dependent on economic motives. It reflects the distribution and redistribution of economic opportunities among regions and countries. The decisions of migrants are based on comparison of opportunities in places of origin and destination (Westin, 1996, 1999).

Another type of voluntary migration is chain migration, which often involves migration to third countries. Chain migration refers to the common patterns when people migrate to a particular destination in which they already have relatives or friends who once have emigrated from the same area of origin. Nowadays, this type of migration is very common among refugee populations (Hassanen, 2000). According to Akuei (2005) migration to third countries takes place under one or more of the following conditions, namely, when: (i) Conditions in the first countries of asylum are unfavourable; (ii) Refugees do not feel safe in first countries of asylum; (iii) Their human rights are violated; (iv) Host governments' policies are hostile to local integration; and (v) Refugees do not want to return to their countries of origin (Akuei, 2004, p. 3). Additional to this is the desire of refugees to be resettled in one of the countries of north where they may have their relatives, friends or neighbours who either finance their illegal journeys or help them to emigrate under family reunification programmes. Chain migration and migration to a third country are interlinked and much of the literature on migration points out that these types of migration are dominant among refugee communities.

As pointed out in the introductory chapter the group of people I have studied left Eritrea against their will. Their lives are not in danger, as they were before. If these people want to return they can decide when and how to do so. In their previous flight situation they did not have the chance to do so.

Forced Migration

The history of Africa is marked by exploitation at the hands of colonial powers. Many historians and geographers argue that the colonial borders were drawn without the slightest consideration of the wishes of the inhabitants of Africa. These borders were created to suit the interests of the colonial powers and, as a result, many of the present borders are unstable (Westin, 1999).

Refugee studies (Moussa, 1995; Sorenson, 1990) show that in the last four decades the Horn of Africa has been marked by wars and conflicts that have forced citizens to flee in search of international protection. During the Derg's rule,[1] Ethiopia was one of the major refugee producing countries in the world (Kibreab, 1987). It was also a major refugee-hosting country. The refugee problem in Sudan dates back to the mid-1950s as a result of the civil war that broke out in Southern Sudan. This conflict was brought to a peaceful ending when a peace agreement was signed in Addis Ababa in 1972; this peace lasted until another round of bloody civil war again broke out in 1983.

Refugees are victims of human rights violations in their countries of origin and thus leave in order to seek protection elsewhere. According to the international instruments relating to refugee status, a refugee is one who flees due to well-founded fear of persecution based on religion, nationality, political opinion, or membership in a social group (Moussa, 1993). The institution of asylum was first developed in Europe necessitated by the need to provide protection to those who were displaced because of occupation and shifted borders after World Wars I and II. In developing countries the refugee problem was the result of the conflicts and disorder that accompanied the process of decolonization and state building (Westin, 1996, 1999).

Although most refugee situations are the result of human rights violations and conflicts, it is important to realize that people flee, not for one single reason but for multiple and inextricably linked political, economic, social, and environmental reasons (Kibreab, 1996). In countries ruled by dictatorial regimes, people often feel insecure and may flee in search of freedom and safety. It is important to observe, however, that although the initial reason for displacement is insecurity, once refugees reach safety, they ceaselessly seek opportunities either to establish themselves in host societies or to become part of the transnational global communities.

Refugees flee their areas of origin in order to seek protection elsewhere. However, not only are the countries where they seek protection in many cases themselves violators of their citizens' human rights, but they are also very poor to the extent of not being able to provide for the basic needs of their own citizens. Being a refugee in a country where there is lack of human rights means that one may continue to be subjected to mistreatment or abuse. In many African states, refugees suffer restrictions of their freedom of movement and residence, employment and access to social and physical services (Kibreab, 1996c; Fellesson, 2003; Horst, 2003). The restrictions are deliberately imposed to prevent integration of refugees rather than to facilitate their integration (Crisp, 2002). Most host governments are opposed to self-settlement of refugees and therefore prefer to place them in government-designated locations. For example, Sudan places Eritrean refugees in reception centres, land-based settlements and wage-earning settlements (Kibreab, 1987, 1990). Those who are caught residing outside of such camps and settlements are punished by law (Karadawi, 1999). Most of the restrictions imposed on refugees' rights in host countries are contrary to the principles embodied in international refugee law and international human rights law. However, violation of refugee rights does not only occur in Africa but is becoming increasingly common even in northern countries (Westin, 1999; Crisp, 2002).

Despite local restrictions, refugees find ways of circumventing such restrictions. Those who have members of their ethnic, religious or clan groups in receiving countries usually find it easier to evade the restrictions imposed by host countries than those who lack such links (Horst, 2003a; Fellesson, 2003). A considerable proportion of Eritreans in Sudan belong to the later and thus live outside of camps and settlements (Kulhman 1994; Kok, 1989; Bascom, 1989; Wijbrandi, 1986; Kibreab, 1996b). Those who share common ethnicity and religion with the host populations can easily blend into the greater society, because they share common norms and values with the host society. Moreover, according to UNHCR 2002 memo such attributes among refugees' communities affect the decision making regarding return migration. Refugees in Kassala belong to this category of self-settled refugees.

Return Migration

In recent years the issue of return migration has attracted the attention of many scholars in refugee studies. An overarching assumption that underpins most of the literature is that once the conditions that forced people to flee are eliminated, refugees return home (see Black and Koser 1999; Mcspadden, 2000; Ghanem, 2003; Rogge, 1994; Kibreab, 1996b). Although, return migration has been seen as one of the durable solutions, studies show that refugees who spend prolonged periods in exile undergo some fundamental changes and so do the places and the peoples they once left behind (Kibreab 2002; Hammond, 1999; Black and Koser, 1999; Graham and Khosravi, 1997; Ghanem, 2003). As a result, returning refugees may be alienated due to the fact that they have not shared the experiences and the developments that will have taken place after they fled from their country of origin. Moreover, the refugees' lives might have been shaped by the experience of exile. According to Gaston (2005), many refugees who return to their countries of origin experience a degree of hostility partly because they do not share the experiences of those who remained behind and the latter feared that the returnees would compete with them for scarce resources, such as water, arable land, grazing land, employment, school places and health care.

In a volume edited by Black and Koser, the contributors deal with return migration operations in different parts of the world. A common theme that runs through most of the contributions is the notion of home and how the meanings and values that refugees attach to it change over time. The findings of the some of the contributions show that whether refugees return home or not is dependent on many other factors besides cessation of the factors that prompted displacement.

Mcspadden[2] is one of the authors who have studied the question of repatriation of Eritrean refugees. In her book *Negotiating Return* (2000) she critically analyses the negotiations that took place between the Eritrean authorities and the UNHCR staff. Her study was based on interviews conducted with Eritrean authorities and the UNHCR. She analyses the identity of the returning refugees and issue of governance in Eritrea. The importance of her study lies in the fact that she documents the disagreements that marked the negotiations that took place between the government of Eritrea and the UNHCR. Her study also shows that the refugees were completely excluded from the negotiation process that took place behind closed doors. She argues that repatriation is a social process marked by many difficulties and challenges. Faced with a repatriation option, refugees have to deal with a variety of economic and social issues in their country of origin and asylum.

Ghanem's study (2003) on repatriation is unique in the sense that she analyses the notion of repatriation from a psychosocial perspective. The psychosocial approach highlights the fact that the returnee's "home" and belonging do not only change through time, but also according to the different social environments in which s/he finds herself/himself. An individual's mind and behaviour are subject to the influences of the social world around her/him. In the context of returnees and their concept of home, it would hence be a mistake to focus only on the returnees' psyche, or alternatively solely on the social world in which one is embedded, as these two elements are in a constant and dynamic dialectic. When a forced migrant settles in a host country, s/he does not enter a vacuum. The same applies when one goes to back to one's country of origin" (Ghanem, 2003, p. 6-7).

Ghanem's (2003) study was carried out under the auspices of the United Nations Research Institute on mass repatriation. It shows that the general conceptions of place, territory and home country are based on European political theory of nationalism. This theory assumes that there is a natural link between people and particular places and therefore social identity is rooted in a particular place – a nation state. According to this conception the world is naturally made up of clearly bounded politico-territorial entities as sovereign states. Thus, according to this theory, repatriation is seen as a means of recouping one's social identity, which only becomes possible by returning to one's place of origin.

In reality, return movements are very complex and may not always be motivated by the desire of people to resume their identities. The dominant assumption about exile and repatriation tends to cloud the judgments of policy-makers and as a result, they disregard the changes that refugees undergo in exile and how this affects their worldviews and objectives (for theoretical and empirical discussions of the changes refugees undergo in exile, see Ghanem 2003).

Refugees' attitudes towards home and repatriation are also affected by what goes on in their countries of origin. If no positive changes are in sight, refugees may be reluctant to consider repatriation as a viable option (Rogge and Akol, 1989; Rogge, 1994; Kibreab, 1996b). Most studies suggest that the experiences and expectations of refugee populations are varied and therefore cannot be generalized. Many authors point out that repatriation does not necessarily mean an end to the refugee cycle, because the concept of 'home' has different meanings and may refer to other things than location.

This indicates that the environment within which refugees find themselves in a host society affects their identity, as well as their attitude. These are the products of social interactions that occur in the context of exile and settlement in a particular place in countries of asylum. These particular places have their own social dynamics, value systems and social norms. Refugees' conceptions of home are very much influenced by these realities. These realities also shape the process of reintegration of returnees in countries of origin. Hammond (1999, p. 227) as well as Rogge (1994) argue that the difficulties that refugees face upon return are mainly due to the changes these refugees have been exposed to in exile.

The way refugees feel about the situation in their countries of origin influences the outcome of the return migration program. Studies on Tigrayan refugees show that refugees do not wait for cessation of the factors that caused their flight or tripartite agreement that takes place between UNHCR, the country of origin, and the host country (Hendrie, 1996).[3] The author stresses that the way refugees react to situations in their country of origin is dependent on their access to information. In her study the author highlights that the will to return or to stay has nothing to do with the economic conditions in either country. This is supported by her study, which shows that the return of about 164,000 refugees in 1987 from Eastern Sudan was to the northern parts of Ethiopia that was controlled by the TPLF.[4] According to the author, the returnees did not receive assistance from UNHCR because the Organization of African Unity and the United Nations that controlled the areas they returned to did not recognize the rebels. The government of Sudan and the Relief Society of Tigray (REST) organized the return operation of Tigryans from Sudan to the rebel-held areas in Tigray.

She points out that the refugees returned to their original country even though the UNHCR neither accepted the move nor economically supported them. This shows that the refugees' decision was probably based on their affinity with the organization that helped them, namely the REST. If the refugees feel some kind of empathy or identification with the regime in their country of origin or with organizations such as the REST they will choose to return because their affinity with that organisation is what gives them a sense of home in their original country. Unlike Kibreab's study my study shows that people's identification with their country of origin is what matters to the decision making concerning return migration.

If conditions in countries of asylum are unfavourable, refugees are likely to return to their countries of origin if they think it is safe to do so. However, for this to happen, they must identify themselves with their country of origin. Return movements may even happen in the absence of political changes that occur in countries of origin. How refugees identify with the authority, people or culture of return is vital to their decision-making concerning return as well as the feeling of home. I agree with Kibreab that time in exile affects the lives of the refugees in both positive and negative ways. Some will choose to stay in the countries of exile if they get the chance.

Experiences presented earlier, which I share with many other refugees, suggest that if refugees are given several alternatives "home" does not always mean the country of origin. What seems to happen is that many refugees come to see themselves as global citizens. Identifying some of the processes and influences that affect the refugees' perception of self and home is one of the concerns addressed in this thesis. Moreover, I agree with the ideas of Mcspadden, Hendrie and others that refugees do not make voluntary moves unless they are certain about the cultural, political or economic conditions in the destination country.

The Meaning of Home

The meaning of home is, to say the least, a contested subject. The concept of home varies between societies, individuals and/or families and therefore it should be understood from a variety of perspectives. It is difficult to give it a universally valid and acceptable interpretation. Notwithstanding this reality, the international community, countries of asylum and origin tend to attribute a rigid meaning to the concept of home.

In Bakewell's words:

> "There is a sedentary bias in the concept of refugee, which implicitly suggests that people belong to a particular location as if by nature. The separation of people from their place forms one aspect of the refugee problem and the restoration of a person to their place through repatriation is often presented as the optimum

> solution. This simplistic narrative of refugees being able to go 'home' is too often employed without a critical analysis of what they conceive to be home and how it has changed since they were forced to leave." (Bakewell, 1999, p. 1)

This rigid perception of the home is the main reason why institutions consider repatriation the most desirable solution.

Repatriation can be the end of one refugee cycle and at the same time the beginning of another. Although returning to one's home country is the favoured option of some refugees, it cannot be said as the UNHCR assumes that voluntary repatriation is the most desirable solution to the refugee problem under all circumstances. The assumption that repatriation represents the most desirable solution is based on the notion of home that is defined from a sedentary perspective. The notion of "home" as a place from which one comes and to which one gradually returns is what return migration is about (Black and Koser, 1999; Ghanem, 2003). However, the concept of "home" is something different from and often more complex than a geographical location. Even if the concept is linked to a geographical location, the latter is not permanent. It can change with time. Places that once were green might change to yellow because of drought. Landscapes may change due to rain or wind erosion or other forms of physical change. Peoples' memories of home can also change over time.

The essentialist conception of home is based on the assumption that return migration represents return not to a country of origin but to the particular place within the country of origin from which one fled, namely the actual house and land one had left behind (Warner, 1994). To some refugees home may have a cultural and spiritual meaning symbolising the beginning of a new life that is difficult to establish because of the changes that might have taken place in the physical, social and economic environment in areas of return (Black and Koser, 1999), in our case Eritrea. The notion of "home" is defined in several different ways, not only on a geographically based view but also from the psychosocial view (Black and Koser, 1999; Hammond, 1999; Ghanem, 2003; Alimia, 2004). There is a common assumption that at the end of a conflict, a return to one's place of origin called "home" is both desirable and is even referred to by the UNHCR as the 'most desirable solution to the refugee problem.' However, the question which people don't often ask is: Who considers this solution as the best option? The refugees concerned? Governments in refugees' countries of origin? Host governments? UNHCR or Aid agencies? Most people tend to talk about best solutions without bothering to ask the refugees for whom such solutions are said to represent 'best solutions.' Inasmuch as there are multiple causes of population displacement, the solutions to the problem of involuntarily displaced populations are varied and many. It is therefore important to question this conventional wisdom. The notion of home has different meanings to different people.

Contrary to the conventional viewpoint in which return migration is perceived as constituting an end to a refugee cycle, it can also be a beginning of a new life that begins after return. For those who lived in exile for a prolonged period, the process of rebuilding new livelihood systems and communities could be as difficult as in exile. For those who were born and grew up in exile, return may represent a new form of uprooting and establishment in an unfamiliar and unknown social and physical environment. The process of settling in such an unfamiliar environment in the absence of proper support systems could sometimes be as traumatising as displacement (Rogge, 1994; Bakewell, 1999).

The dominant essentialist perspective in the discussion of return migration is based on the assumption that once refugees return home, they recoup their citizenship rights, including their homes, lands and other possessions and consequently become reintegrated into their old communities. This assumption is an oversimplified presentation of a very complex social reality. Neither the refugees nor the populations that have stayed behind are any longer what they used to be (Kibreab, 2002). Everything has to be renegotiated anew. For example, some of the Eritrean refugees in Eastern Sudan have been in exile for nearly forty years. They have undergone some fundamental transformations in exile (ibid). The communities they left behind have also undergone change and more importantly, their areas of origin are now inhabited by various other groups from different parts of Eritrea. Nothing was at a standstill during the years they have been in exile. It is not surprising therefore that those who return have to start anew.

The attitude of refugees towards their homelands is also a function of particular life experiences prior to displacement. It is argued here that the meaning and value refugees attach to their homes of origin is to some extent shaped by their pre-flight experiences. For example, the Assyrians who fled from Iraq after suffering prolonged deprivation of citizenship rights and discrimination do not see repatriation as representing a durable solution to their problem. For this category of refugees repatriation may represent a return to an oppressive regime. In fact, the very notion of return may be loathed as it may rekindle old memories of degrading treatment at the hands of an oppressive regime and hostile communities.

In their study on transnational Iranian communities Graham and Khosravi (1997) argue that exile did not begin with the departure of the refugees from their homeland. Their exile began while they still were in Iran. Therefore, for groups who have had severe and bad experiences before they left, the place of origin to which they are expected to return does not always engender a feeling of joy and homecoming. Home can be a place that rekindles unhappy memories among returnees who prior to their displacement might have suffered degrading treatment at the hands of an enemy.

Refugees are not a homogeneous social group. They are people with different capacities and ambitions. Therefore, their links or commitments to their countries of origin vary. All refugees do not treasure their home countries in the same way. Some refugees' pre-flight life experiences might have been unpleasant. People in this predicament may therefore get on with their lives in the host country without being pre-occupied with issues of return. However, as research on returnees shows, all refugees who wish to return often have to adapt to new social and cultural settings, create new social networks; find new jobs and education for their children (Akol and Rogge, 1989; Rogge, 1994; Fellesson, 2003).

INTEGRATION OF REFUGEES IN HOST SOCIETIES

Although assimilation rather than integration is the term used in the UNHCR's statute, the term integration rather than assimilation is most often used in the literature. The concept of integration is not easy to define and several authors tend to define it operationally depending on the purpose of their studies. One of the best definitions known to the author is Tom Kuhlman's in which he defines the term as follows:

> "If refugees are able to participate in the host economy in ways commensurate with their skills and compatible with their cultural values, if they attain a standard of living which satisfies culturally determined minimum requirements (standard of living is taken here as meaning not only income from economic activities, but also access to amenities such as housing, public utilities, health services, and education) if the socio cultural change they undergo permits them to maintain, if standard of living and economic opportunities for members of the host society have not deteriorated due to the influx of refugees, if the friction between the host population and refugees is not worse than within the host population itself, and if the refugees do not

> encounter more discrimination than exists between groups previously settled within the host society: Then refugees are truly integrated." (Kuhlman, 1991, p. 7)

As said earlier, integration into first countries of asylum is considered as one of the solutions to the refugee problem (Kuhlman, 1994; Wijbrandi, 1986). The 1950 Statute of UNHCR mandates the organisation to seek solutions to the refugee problem in collaboration with governments in countries of asylum. One of the solutions envisioned by the Statute is assimilation in new communities. However, the policies of host governments in developing countries are based on local settlement rather than on local integration, that is they offer settlement in isolated camps but not among the local people (Kibreab, 1987). In fact, as Kibreab argues, such settlement policies are designed to perpetuate refugee status by preventing their integration into host societies (ibid.). The local integration option promoted by donors, UNHCR and NGOs is generally rejected by host governments in developing countries (Fellesson, 2003; Kibreab, 1987, 1996; Crisp, 2002; Kok, 1989).

Integration does not imply assimilation of refugees into host societies. Assimilation implies relinquishment by immigrant or refugee communities of their culture and way of life in favour of adapting to the cultures and lifestyles of host societies and over time become indistinguishable from members of the dominant culture of the host society. Integration on the other hand, is about maintaining one's own identity and still becoming part of the host society to the extent that refugees and host populations can live together harmoniously (Crisp, 2004, p. 2). Integration refers to a process in which a refugee or migrant community becomes part of the greater host society without abandoning its culture and way of life. It is in this sense the Statute of the UNHCR and the 1951 UN Convention Relating to the Status of Refugees conceive solutions to the refugee problem within the context of first countries of asylum.

Kunz (1979) identifies different factors that are related to integration of refugees in host societies. These are home area related factors, host country related factors and migration related factors. The home area related factors are about the situation of the refugees in their country of origin. Host related and migration related factors are about the willingness of the refugees to settle and the willingness of host societies to accept refugees, as well as about availability of resources essential for livelihoods. Bulcha (1988) measures social integration by the extent to which individuals are linked to social groups who are within their social environment. He argues that individuals do not exist by themselves and therefore they are not separate from society. Fellesson (2003) regards integration as a link between individuals and society reflected in commonly held values. Social bonds that emerge from adherence to similar values and norms may be understood in terms of solidarity. Fellesson also points out that integration has cultural and social aspects. The cultural aspects include learning, which is practice reflected in becoming acquainted with cultural values and norms of a host society while maintaining one's own cultural values, norms and way of life.

Political integration is based on the possibility to participate in the political life of host societies. Political integration is only feasible after legal integration resulting in naturalization. This is non-existent in nearly all the developing host countries. Crisp (2003) points out that most refugees in a protracted situation will eventually become integrated to their host society through the time spent in exile. This is, however, dependent on the policies pursued by host governments.

Bulcha (1988) defines economic integration as being synonymous to self-reliance, a status reached when refugees find jobs and become part of the regular labour force. This is on the one hand dependent on whether or not refugees are legally allowed to participate in the labour market and on the other, on demand for labour. The absence of one may prevent refugees from achieving economic integration.

Kuhlman (1990, 1991, 1994) who conducted studies among Eritrean refugees in Kassala and its surrounding area differentiates between integration in multi-ethnic societies and in ethnically homogeneous society. He stresses that in a multi-ethnic host society, refugees can be integrated quite easily because such societies are familiar with socio-cultural differences. He further identifies a number of indicators as a measure of integration. These are: 1) Changes that refugees undergo in their host country; 2) Friction that exists among multi-ethnic groups in a host society; and 3) The degree of experience of discrimination that refugees face in a host society.

Economic integration does not take place in a social vacuum. This requires conscious policy of training and job creation. This also requires well-developed webs of social networks through which refugees can gain information about job and training opportunities, as well initial support. In the case of Eritreans in eastern Sudan, religion and ethnicity play a vital role for employment, access to education and health care facilities as well as for freedom of movement and residence. Those ethnic groups who share common ethnicity, religion and language with sections of the host society have access to the labour market on equal footing with nationals. Those who lack such connections experience discrimination and exclusion on a daily basis.

Although Bulcha and Kuhlman note that ethnicity, shared values and norms are factors that influence integration in the host country; the authors do not go deeper into the subject. Unlike them I try in this study to see these norms and values from a deeper perspective. As I did, Kuhlman and Bulcha carried out their studies in the Kassala region. In this area norms and values are by and large shaped by Islamic culture. Unlike them however, I tried in this study to see the way in which Islam guides its supporter, and out of that, I seek the link that connects the Moslem refugees with the local people.

The Role of Remittances and New Technology for Survival

√ Recent studies about communities in Diaspora have shown the significance of remittance sending. Remittances from relatives in transnational Eritrean communities are a way to secure means of livelihood for many Eritrean refugee families in Kassala. Remittances include cash sums regularly sent by thousands of Eritrean refugees and migrants back to their relatives in Kassala town (Kibreab, 1996b; Al-sharmani, 2004). On the one hand, these remittances enable such families to survive in Sudan and on the other they create a positive image about the countries from which the remittances are sent. The positive role of remittances is very broad. Today remittances make up a significant part of the international financial flows to developing countries; only a small proportion is invested in business or production. Remittances in the form of cash have been seen as one of the major benefits of emigration for developing countries (Sorensen, 2005).

√ Since remittances are largely directed towards migrants' families or communities of origin, they can increase inequalities, as individual families are the primary beneficiaries. Migration does not occur randomly with people drawn evenly from a given country. Migrants tend to follow paths established by family members, friends or neighbours. As a result, some communities may receive remittances from numerous relatives overseas, whilst others receive nothing at all. It is also important to note that migration is an expensive undertaking requiring significant investments and thus, is only available to those who have sufficient resources to draw on. The poorest of the poor are less likely to migrate or receive remittances. The majority of transfers are made in small amounts to help support household budgets and to fund children's education, healthcare and housing. In conflict-affected areas, the financial support from members of the Diasporas is essential for poor households' survival and post-war recovery. Where families incur debts to fund migrants' journeys by paying extortionate fees to human smugglers, remittances can help repayments of such debts.

√ Another factor that facilitates the connections and helps families in Kassala to have contact with their relatives in Diaspora is the new technology. Refugee studies show (Horst, 2003b) that refugees maintain links with their relatives who are settled in the first country of asylum through a complex network. The advances in transport and communication technology, especially, the Internet and radio transmitters have facilitated the maintenance of contacts over long distances (ibid.). These transnational relationships are not only focused on the links between migrants and their families but these relationships also foster links between refugees in the areas of destination.

For example, those who migrate to democratic countries, which operate within a strong framework of law and respect for human rights may return with very different attitudes to their countries of origin. Moreover, returning refugees can play a vital role in the reconstruction of the country and economic recovery. However, the highly politicized and often insecure environment and weak economic base in many countries affected by widespread conflicts make repatriation not a straightforward option. Large-scale repatriation can put enormous pressures on already stretched resources. Moreover, repatriation of refugees from Europe may also cut off the remittances, which may play a vital role in post-war recovery (Van Hear, 2003).

As pointed out above money transfer and new technology, and respect for human rights issues are associated with many encouraging aspects. This is why refugees in Sudan wish to resettle in third countries in the north, and this is also why these two factors play a great role for the decision making regarding return migration. This study aims to understand how the presently mentioned procedures affect the decision-making of the refugees in Kassala concerning return migration.

CHANGE AND CONTINUITY: GENDER RELATIONS

Research on international migration rarely considers the role of gender in migration despite the fact that men and women play different roles in the migration process, face different difficulties and opportunities to migrate and are affected differently by displacement. In contrast to sex, which is biologically determined, gender is a social and cultural construction (Silberschmidt, 1999; Duncan, 1994; Dolyal, 1996). The concept of gender was first used in feminist literature in the 1980s. Moreover, gender relations refer to power relations that exist between men and women. These power relations are prescribed by gender roles. Furthermore, feminist research interprets gender relations as men and women being embedded in a symbolic and normative division. Femininity is explained in terms of the other. This is to say that masculinity serves as the standard for defining human beings in society. What a woman does, what she thinks, and how she defines herself is seen from standards set by masculinity, which takes 'the man' as the norm and consequently the woman becomes the other (Lutz, 1991, p. 2).

The available literature indicates that migration in the 1960s and 1970s was gender-specific. The large majority of migrants were men. However, recent studies show that female migration has become a significant part of voluntary and involuntary migration. Women from poor countries are migrating to rich countries as labourers. A case in point is the migration of Mexican women to the United States of America and of Eritrean women to the

Middle East and West European countries. In the past, it was assumed that women and children stayed behind whilst men migrated to cities in search of employment. Nevertheless, Gilbert and Gugler (1982) point out that this idea does not hold for all developing countries. In fact, women in Latin America and in Asia are predominantly living in cities, not in rural areas, which, among other things, is due to local and national traditions, as well as to severe shortages of resources, including land.

Female migration is associated with the new global economic transformation and restructuring of labour migration. This means a new group of immigrants is on the rise. This takes the form of single women and female family breadwinners who move both independently and under the authority of older relatives (Sorensen, 1990).

To study the changed gender roles as a result of migration is important. Migration affects gender roles. Women who used to be excluded from the public sphere started to participate in the public sphere as a result of migration (Kibreab, 1992; Connell, 1998). These studies show that women send back home a greater share of their income in remittances in comparison to men. They also save a greater proportion of their incomes (Sorensen, 2005). Such important findings would have been lost had it not been for studies that are gender-centric. Studying migrants without including the gender aspects exposes us to the possibility of not being familiar with vital issues that affect lives of migrant women.

Notes:

1. The government that ruled Ethiopia after the fall of the King of Ethiopia.1974-1975.

2. Lucia Ann Mcspadden is Ph.D. in cultural anthropology and researcher at the Life and Peace Institute, Uppsala, Sweden. Dr. Mcspadden carried out research on the resettlement of Ethiopian and Eritrean refugees in North America as well as in Eritrea.

3. Barbara Hendrie is a social anthropologist based at the department of anthropology at the university collage of London. Her study is about the repatriation of Ethiopian refugees from Sudan to Ethiopia. She lived in Sudan from 1984 to 1988, and she worked with NGOs that acted as a channel for relief assistance to non-government controlled areas of Ethiopia and Eritrea. The author describes what happens when a repatriation scheme of refugees is implemented and how it was decided whether it was safe to return or not.

4. Tigray People's Liberation Front.

3

METHOD, DATA SOURCES AND PROCESSING

In this chapter I shall discuss the case study approach, language issues in the fieldwork and research ethics. An attempt will also be made to present the method of data analysis. The chapter will conclude by making some remarks on the validity and reliability of the data.

Qualitative Method

There is a body of literature dealing with participatory observation and interviews (Kvale, 1996; Limb and Dwyer, 2001; Ramazanoglu and Holland, 2002). Qualitative methods give a researcher an opportunity to be in the "world" of the people who are studied. One of the important aspects of qualitative methodology is that it does not entail pre-existing fixed assumptions about the social world. Thus, the aim here is to understand and explain the social world and everyday life of the refugee communities through observation, listening and reflecting on their experiences and realities (Dwyer and Melani, 2001, p. 6). This is because qualitative methodology is characterized by an intensive search for meaning rather than numerical descriptions. Researchers who use this method seek subjective understanding of social reality rather than statistical description or generalized statements. The most suitable method for the data collection of this study is qualitative because of the aim, research questions, living conditions of the study group and lack of statistical data. The focus of the chapter is on the means of assembling data as well as on the relevance of the methods chosen and the problems and challenges faced in the process of doing this research.

The literature on qualitative research and personal interviews is broad (see Kvale, 1996; Limb and Dwyer, 2001; Ramazanoglu and Holland, 2002). As much as possible, an attempt has been made to create an opportunity for the subjects of the study to participate in the research process so that the final product gives voice to the refugees who have been voiceless in terms of policy.

Although quantitative methods emphasize the need for objectivity, in reality this is easier said than done. This is because the ways in which researchers select the objects of their studies, the ways they frame their research questions and design their research instruments are to some extent influenced by bias. It is therefore important to acknowledge this from the outset rather than to give the false impression of objectivity.

The choice of methods for a particular research project is determined by how a researcher understands the social reality to be studied. Therefore, how a researcher prioritizes a particular subject of study, frames his/her research questions and what s/he seeks to understand determines the design of the study and the research methods used.

In my case, it cannot be denied that I am connected to the subjects of my study in many ways, including being a refugee, living in a distant place from where I was born.

It is not due to coincidence that I have chosen to write my thesis on this particular group of refugees. My decision to write about this group is intentional and is motivated by the desire to understand their situation, as well as the transformations and changes they have undergone in exile, as has been the case with me and other fellow Eritreans who found asylum and hospitality in Sweden and elsewhere in countries of the north.

The methods employed in this research are interviews and participant observation. This is one way of giving participants an opportunity to be heard in the research process (Kvale, 1996). Talking to the subjects of a study enables a researcher to know about how they view and understand their world (ibid). The best way to know their social world is by being close to them.

Case studies provide knowledge, insights and information about the group that is studied and their surroundings. By applying, the case study method a researcher seeks to uncover life experiences and to document information about the subjects of the study (ibid.). The data gathered in this manner provide the researcher with insights into the past and present lives of the subjects of study. By comparing the past with the present, it becomes possible to obtain a picture of the real world of the subjects of the study (Kvale, 1996; Nachmias and Nachmias, 1996). It could have been helpful to use statistics as a complementary source of information. However, in the field that I studied there were no accurate or reliable statistical data to access.

The use of the case study approach is widespread both in countries of the south and north. This study draws its methodological inspiration from the works of Kvale (1996), Nieswiadomy (1998), Ramazanoglu and Holland (2002), Limb and Dwyer (1999), Nieswiadomy (1998), Pile (1998), Strauss and Corbin (1990) and Eyles and Smith (1988). The common theme of these studies is that they all value the importance of qualitative methods such as participant observation and interviewing as a means of data collection.

The work of all these scholars, with the exception of Pile (1998), is concerned with conditions in societies of the north. It is still possible to draw some useful insights that can be applied to research on societies of countries of south as in my own case study. One other piece that has substantially influenced my approach and perspective is Robina Mohammed's (2001) work on gender relations in south Asian diasporic communities.

To get an idea about the social reality of the Eritrean refugees in Kassala town, the methodology to be employed in the study should be one, which enables me to understand the life world of the Eritrean refugees from their own perspective. These data are therefore gathered through interviews and participant observation. This method of data gathering was also selected taking into account the financial and time constraints within which I operated.

The main fieldwork was preceded by an extensive review of the available literature. Given the paucity of data sources in Swedish libraries, the large majority of the useful literature was obtained from the Refugee Studies Centre at Queen Elizabeth House, Oxford University, the British Library, and the School of Oriental and African Studies (SOAS) in London. Some useful secondary sources were also collected from the university libraries in Khartoum, particularly the Ahfad University for Women. Some useful primary data were also collected from the archives of the Commission of Refugees in Sudan (COR).

The Fieldwork

Research Permit

In Sudan it is impossible to conduct research without a research permit. The government is reluctant to issue research permits in areas where security is a major concern. Since Kassala town is close to the Eritrean border, security is a major concern for the government.[1] This was further complicated by my Eritrean origin. Thus, the process of obtaining a research permit was both difficult and protracted. My application for a research permit was filed to the Office of the Commissioner for Refugees (COR) in Khartoum. COR wrote a letter of introduction to the security police in Khartoum. After that I was interviewed by the security police about the purpose of the study, its duration and how the results would be utilised. The security police would only issue me a research permit on condition that I submitted a copy of the findings to them upon the completion of the thesis. After receiving clearance from the security office, COR wrote a new letter to the regional manager at Showak.[2] The Showak office in return wrote a letter to the COR office in Kassala town. I submitted the letter to the COR's office in Kassala. The head of COR in Kassala sent me to the security office where I was interviewed again and asked about the aim and purposes of the study, as well as my own background. After completing these protracted procedures, I was free to move and carry on with my work in the town. I asked COR officials whether I could take photos and I was told that a permit was necessary for this. My application for this was turned down and I was unable to take photos of the respondents and their families, nor of their work places.

Two Factors that Contributed to the Situation of this Study

The two facts have especially contributed to forming the conditions of this study: First, at the time of my fieldwork the UNHCR announced the cessation of refugee status for all Eritrean refugees worldwide (the announcement was meant until end of December 2002). This gave me the opportunity to collect information about the campaigns arranged by the UNHCR. Moreover, information was collected for different refugee groups through individual discussions, informal meeting and group discussions. I have used this information as supplementary to the data from the key informants. Secondly, my personal acquaintance and knowledge of the socio-cultural conventions of the host country and refugees was of great importance for my access to the field.

Formally, Eritrean refugees are not allowed to settle in urban areas, which includes Kassala. As a result, they are invisible in all official records of Kassala town. My entry into the 'invisible world of the Eritrean refugees' was greatly facilitated by my insiderness, which was reinforced by my intimate knowledge of the local languages and cultures of the refugees in Kassala town. Not only do I share the religion of the majority of the refugees in Kassala town – Islam – but I also belong to one of the ethnic groups covered in the study, namely the Blin that has a long-standing link with the majority group among the refugees – the Tigre. I am fluent in Tigre, Blin, Tigrinya and Arabic. These are the major languages of the refugees in Kassala town. My familiarity with the cultures, way of life, Moslem religion, ethnicity and languages enabled me to collect data without intermediaries or interpreters.

This thesis is thus based on diverse sources. The main data are from personal interviews conducted with 24 refugee informants, comprising of thirteen women and eleven men. All of the key informants were from Kassala town. The interviews were conducted in Arabic, Tigre, Tigrinya and Blin. On top of the fieldwork conducted between August and November in 2002, two other shorter fieldworks were undertaken in 2001 and 2003. The 2001 fieldwork was exploratory and was helpful in preparing the ground for the main fieldwork. The 2003 fieldwork was undertaken to supplement and update the data collected in 2002. During the three months of fieldwork, I lived with the refugee families in Kassala town.

Participant Observation and Informal Interviews

Besides the 24 individual interviews, participant observation represented an important additional method of data gathering. Participation in the meeting organized by the UNHCR, attended by a large number of refugees, was an important source of information. I also visited many families and community leaders and had discussions with them about return to Eritrea. I was able to gain some critical insights and knowledge, which were helpful in supplementing the data I gathered from key informants.

The UNHCR office in Kassala represented the first entry point to the whole fieldwork process. Since the UNHCR is the main body with a mandate to provide international protection and assistance to refugees, I expected to find some useful data about return migration. However, the focus of the office at that time was on repatriation of Eritrean refugees. They gave me information pamphlets they had prepared to inform the refugees about the conditions and the kind of assistance and protection they would receive if they return to Eritrea.

I was also informed about the forthcoming information dissemination meeting in Kassala town. The organisers kindly invited me to attend the meeting so that I could learn how the process of informing the refugees about the conditions in Eritrea as well as UNHCR's responsibilities of protection and assistance operates when the refugees return to Eritrea. I attended the meeting as an observer and this gave me an important insight into how repatriation operations are initiated.

Since this event represents an important aspect of the refugees' decision concerning return to Eritrea, it was a valuable undertaking on my part. Of particular interest were the questions the refugees asked. Some of the questions were very informative and opened a new avenue of inquiry and enabled me to understand some important questions, which I was unaware of before. The questions focused on issues pertaining to the political situation, livelihood issues, education, national service and imprisonment of journalists and religious leaders, and others in Eritrea. These questions were not raised in search of answers from the UNHCR but rather to question the appropriateness of the planned repatriation operation to a country where, in their view, violations of human rights were ubiquitous. In other words, they were directly or indirectly telling the UNHCR that although the conditions in connection to their flight from Eritrea had ceased when Ethiopian rule was thrown out of the country, the prevailing conditions in the post-independence period were still unfavourable for repatriation.

When I returned home in the evenings, I discussed the issues that I heard being raised in the meetings with relatives, acquaintances and neighbours. I also conducted similar discussions in the market place with members of the Eritrean refugee communities. The data and the insights that emerged from these discussions were invaluable. Before entering the field, I had a theory in mind that I thought would serve as a tool of analysis and explanation regarding the factors that would influence the refugees' decision concerning return migration. My belief was that when the factors that caused the flight came to an end the refugees would return to Eritrea. That is their feeling of home should be associated with the political change in Eritrea. However, the reality in the field was different from what I assumed. Thus, the theory I had in mind proved inadequate. The reality on the ground was so complex and continuously changing that there was a needed for a new theoretical framework, which allowed flexibility. A shift of perspective was also needed to understand the social, economic, as well as political complexities within which the refugees made their decisions concerning return migration.

In each house that I visited in Kassala town,[3] I met many Eritreans. A group discussion was held with female household heads who had previously returned to Eritrea in response to the political changes that took place with Eritrea's independence, but who had subsequently returned to Sudan. The information gathered from this discussion was very useful. Invaluable information was also collected from those who worked in the repatriation registration centres, and the refugee committee (Legina*)* in the refugee camps of Wedsherifey and Girba. I personally visited the refugee camps of Shegerab and I had the opportunity to interview members of Legina at the two camps in Kassala. I also attended meetings in which the UNHCR, COR and ERREC jointly held to inform the refugees about the changes that had taken place in Eritrea and to tell them that the time had come for them to return to Eritrea. Three such meetings were held in Kassala town, two in Khartoum, two in Kashm el Girba, one Umagata and one in Kilo five. Attending all these meetings helped me to gain some critical insights. My attendance in the repatriation campaign was very useful and as a result, I gained a deeper understanding of the complexity of the decision concerning repatriation. The overall issue of repatriation, as well as the appropriateness of the actual time of return was openly discussed by the refugees. Some of the refugee leaders were against repatriation and they spelled out the reasons why they thought so. This gave me an opportunity to understand the arguments of the refugees. Of course, in meetings such as this, it is the articulate and the loud few that dominate the scene.

One of the issues the study tried to explore was the extent to which the attitude towards repatriation varied between the first and second generations. In Kassala town, special separate focus group discussions were therefore held on the one hand with a group representing those who came to Sudan as adults and on the other with a group representing the second-generation refugees. The same procedure was replicated in Kashm el Girba refugee camp. This time the participants in the group discussions were women representing the first and second generations. In Khartoum, informal conversations were conducted with young women and men who worked in Sudanese restaurants. Invaluable data were elicited from these informal encounters. On top of the data gathered from the above named sources some useful information was gathered from informal discussions held with group of refugees (from the Nara clan) residing in the El-Sebil district, to get an idea about what was going on regarding return migration in Kassala.

Since I lived among and with refugee families, a lot of insights and knowledge were gained through the method of participant observation. Some data that I used for comparison in the analysis were also collected through informal discussions with Sudanese informants in the parts of the city where there were large concentrations of Eritrean refugees. This process of data collection was also facilitated by my local knowledge but more importantly by my fluency in Arabic, as well as familiarity with cultural practices of the local inhabitants of Kassala town.

The fieldwork and the research process for my Ph.D. thesis was demanding, challenging and distressing, at times even frightening, but also exciting and enriching. On the one hand, the research experience enhanced my knowledge about the people whose identity, history and culture I once shared, and on the other hand, the endless conversations with key informants rekindled some unhappy memories and trauma that commonly accompany the refugee experience.

Kassala town has grown massively over the last few decades. This is partly due to natural population growth but also as the regional capital located in the central rainlands of Sudan known as the 'bread basket' of the Middle East, the town receives large numbers of immigrants from Western Sudan, West African countries and also from Eritrea. The Eritrean refugees are mainly concentrated in three parts of the town, which has considerably facilitated my work. The large majority of Eritrean refugees are concentrated to Muraba'at, Abukamsa and Souq (Kassala market). The first two are residential parts of the town, but the third is the hub of all forms of economic activities, especially trade. If the Eritrean refugees had been scattered throughout the town, my task would have been very difficult if not impossible. Although I learned a lot in the campaigns, I had to reach the silent refugees by arranging individual interviews.

Formal Interviews

I was once a refugee in Kassala myself and thus familiar with the social environment within which the research was conducted. Besides the fieldwork carried out in 2001, 2002 and 2003 I had also previously conducted survey research in 1996 on female genital mutilation among refugees in Kassala, which provided an important background knowledge and experience to this study. Over time, I have developed a rapport with some members of the refugee communities. I had also developed a link with a group of informants comprising of relatives, acquaintances, their neighbours and others in Kassala town. These connections and networks were significant, not only for providing entry for the fieldwork but they were also important sources of information.

The snowball technique was used as a means of identifying the various interviewees for the study. According to Altamirano (2000) and Jacobsen and Landau (2003) the snowball technique of sampling is a way of selecting respondents and acquiring information as the research proceeds, rather than using an already defined sample from the outset. It is through contacts with informants obtained from interviews that new respondents are found. The weakness of snowball sampling is that some biases may arise because of the limited scope of the social networks. In this particular case study, there was no alternative to the snowball technique of identifying respondents.

My old contacts served as my entry points. I asked the people I knew to introduce me to the groups or individuals I did not know from before. I sat down with the people I knew and prepared a long list of people who met the criteria of being included in the list of potential interviewees. One of the most important approaches I adopted was the principle of inclusion. For the study to reflect the reality of the refugees in Kassala town, it was important to include individuals from the different social backgrounds (see below).

The snowball technique of identifying individuals for the interviews required the co-operation of the refugees, particularly of those who fled Eritrea in the post-independence period. Identifying interviewees from this category was only possible with the assistance of relatives, friends and neighbours. In order to minimize the risk of bias towards certain categories of the refugee groups, I tried to adopt multiple entry points for the snowball sample. This helped in broadening the scope of the interviewees. The interviewees were selected from three different parts of the town (Muraba'at, Abukamsa and Souq) where Eritrean refugees were concentrated. This selection was aimed to reduce the potential for bias.

In 2001 I prepared an interview guide for the pilot study. It focused on the general factors that influenced the decision of the refugees in the camps to repatriate to Eritrea or stay in Sudan. The results of the 2001 pilot study were very important in terms of providing knowledge that enabled me to change, modify and remove some of the questions. It also helped me to identify gaps that required to be filled. Hence, new questions whose importance was not seen in the pilot study were added. So before I started the fieldwork in 2002, the research guide was considerably revised in accordance with the results of the pilot study. The research guide or the interview schedule was semi-structured and open-ended. This was intended to give an opportunity for further probing as new issues arose during the conversations with respondents. The research guide was applied flexibly.

The interviewees were selected to represent the refugee populations by taking into consideration sex, age, household headship, ethnicity, religion, education and marital status. Of the thirteen women four were university graduates, three were high school graduates, three were literate in Arabic and three were illiterate. Seven of the male respondents were university graduates, two were junior high school graduates and two were senior high school graduates. One male respondent was illiterate. Of the eleven men only two fled from Eritrea after independence, while the rest had lived in Sudan for more than 20 years. Of the eleven men, seven were active members of the ELF while three were ex-members of ELF and one was an ex-member of the EPLF.

Of the thirteen women, ten had lived in Sudan for more than 20 years while three had fled Eritrea after independence. Out of the thirteen women, only six were active members of ELF while three were ex-members of the ELF. Although an effort was made to cover all sections of the refugee communities in Kassala town, it is difficult to claim that this is a representative example of the population. Many of the refugees, particularly those who fled Eritrea in the post-independence period lived hidden lives and therefore not only were their numbers unknown but they were also reluctant to come in the open.

Most interviewees sounded unsure in the beginning, but their suspicion was overcome gradually. After being introduced to the informants, my next step was to win their trust. Once I introduced myself and described my background, as well as my link with Kassala town, they seemed to like the idea that a person who was linked to them by ethnicity, religion and country of origin was en route to become a 'doctor' as they put it themselves. Some of the respondents had already heard about me and some even knew of my family and all this contributed to my efforts to win their trust. I told every interviewee, that s/he could interrupt the discussion at anytime s/he felt uncomfortable. However, this never happened.

Most of the interviews were conducted at places chosen by informants, namely in shops, in the shade of trees, inside homes and in compounds. Most of the recorded interviews took place in the homes of informants. Some interviews took longer to conduct than others. The interviews that took longer time to complete did not necessarily produced better quality material than the others. The majority of the recorded interviews took between 40 and 50 minutes. Some of the recorded interviews took between two, and two and a half hours. At the end of every interview, I promised to send to each informant a copy of my thesis if they gave me their addresses. They also told me that they were very happy to give me information that would enable me to write my thesis and to achieve my goal. Several elders said, "You are Eritrean and we want you to succeed and serve your country. Good luck and may Allah be with you." The number of individuals interviewed in Muraba'at was higher because this is the area in Kassala where Eritrean refugees are more concentrated than in the other two parts of the town that were included in the study. Although interviews were conducted with twenty-four key informants, a number of focus group discussions and informal interviews were also held in different settings.

The large majority of key informants in the interviews said that it was the lack of politics of inclusion and democracy that forced them to stay in exile. However, in the informal interviews, most of the interviewees said that Mr. X's family returned to Eritrea but could not stay because life was more difficult there than in Sudan. Ms. Y's family could not stay in Eritrea because her family was not welcomed by their relatives in Eritrea. Some families could not stay in Eritrea because their children were not used to the manner and life-style in that country. Had it not been for the informal and casual interviews, all these important insights would have been missed. However, as pointed out before the information from the key informants are the main data and the rest is used as complementary information. The data from the formal interviews were important, but the information I gathered in unofficial settings over a cup of tea or the usual Eritrean coffee ritual was more frank and in many ways more useful. For example, when asked to explain the reasons why they failed to return home when the political conditions in connection with which they fled Eritrea had come to an end, the tendency was to place greater emphasis on political rather than on economic and social factors.

I also conducted interviews with staff at the Eritrean Embassy in Khartoum. Interviews were also held in Khartoum in September 2002 with members of staff of UNHCR and COR in the registration centres of Demi, Gerf and Bahir.[4]

The 'Insider–Outsider' Divide

Being an insider is to share common cultural, political and historical experiences with the group being studied whilst being an outsider means to lack such common experiences. These terms are important because both have positive and negative consequences for the outcome of a study. Being associated with the subjects of study may normally occasion a positive reception, but this cannot be taken for granted because being an insider can also have unfavourable consequences. For instance, Muhammad (2001) observes:

> "This belonging was seen to endow me with a superior, almost organic knowledge of the community not accessible to outsiders, for example white people" (Mohammed, 2001, p. 102).

In order to be accepted, she had to dress herself in accordance with local custom and use her last name.

Depending on where and when the study concerned is executed, there are certain constraints a female researcher has to deal with that may not apply to her male counterparts. In my case, to be accepted by the refugee community and their local hosts, I had to dress in accordance with the local cultural practices. I had also to behave as any woman in that particular community would. So looking like the informants makes approaching them much easier. My dress and my behaviour helped me to be accepted by members of these the communities.

PROBLEMS ENCOUNTERED DURING THE FIELDWORK

Conducting research, particularly in a country where gender roles are very rigidly defined, is associated with many problems. According to the local norms and values, a woman's proper place is in the home. Since I did not comply with many of the traditional norms and practices, some viewed my undertaking with a certain degree of suspicion. One respondent said, "Why send an Eritrean woman to carry out a study on Eritrean refugees? What are Eritrean men doing in Sweden?"

Political life among Eritrean refugees was at that time highly politicized. Although my case was facilitated by the fact that some of the refugees knew my family, those who did not know my background did not immediately trust me and required assurances about my status. They suspected me of spying for the Eritrean government and that was in Sudan to register their names in order to repatriate them by force. This shows how the refugees were scared of outsiders and how they distrusted the host government's protection. However, I tried to explain to them that I have nothing to do with any authority or NGO; I assured them that I am independent from the authorities in Sudan, Eritrea or UNHCR. I also gave them assurance that their anonymity would be protected and that the information would be treated with utmost confidentiality. Some also thought I was sent by the Sudanese authorities to document their property ownership in Kassala. Thus they were unwilling to admit that they owned any kind of property in the Kassala despite the fact that this was common knowledge among the refugee communities in Kassala town.

As we shall see later, some of the refugees had managed to obtain informal Sudanese citizenship (*jinni*). This was possible for those who have strong contacts and links with the local people. However, refugees do not talk about it openly with outsiders. Acquired citizenship is only informal because Sudan does not have a policy of naturalisation.

There is a tendency among the refugees to associate individuals who come from Europe or North America with assistance providing international organizations (Kibreab, 1987, 1990). Therefore, it was important on my part to dispel this belief; otherwise, the risk of collecting wrong information is high. If the respondents suspected that I represented aid agencies, they might have exaggerated their difficulties in order to maximise receipt of assistance. Before asking any other question, I made sure that the interviewees fully understood that I had nothing to do with aid agencies and they would receive no material gain from the research. It was important for me that to put this message across in unmistakable manner. I told them that the findings of the study will be published and whoever wants to understand their situation will be able to read it. I assured them that no harm or direct benefit would result from it. I also assured every respondent that they will remain anonymous and no one will be able to identify them.

There is a body of literature on the importance of social networks among African refugees. The question of social networks is critical in the livelihood and survival systems of the Eritrean refugees in Kassala town. This is because formally, they are not supposed to be there and therefore they do not receive any assistance from UNHCR. Social networks are therefore very important for their livelihoods and survival. Although most of the refugee respondents were open and forthcoming, there were some questions they did not like and it was clear to me from the reading of the secondary sources (Horst, 2003) that the success or failure of one's study depended not only on asking the right questions at the right time but also on avoiding questions which the refugees regarded as sensitive and private.

There were also other 'no go areas' among the respondents. Asking questions about gender relations and about women's role in the decision concerning return migration was, for example, a proscribed taboo. Although I thought this question was crucial, I did not want to pursue it at the risk of failure. One of the advantages of being an 'insider' is that one is aware when information is unclear to fit into a picture an informant wants to show. I was not in a position to contest what I heard but it was clear to me that such information was of no use for my study. However, a compromise and a trade off were essential in order to access other information, which was revealed by informants because this was not regarded as sensitive.

Another problem I faced in undertaking the interviews was that the refugees believed that every researcher or Eritrean who come from abroad is loaded with cash and would be able to assist them or help them to be resettled in one of the developed countries. Their stories and narratives were therefore shaped by this ultimate goal. Moreover, as an Eritrean I am expected to be understanding, empathetic and helpful.

I was continuously asked by most Eritreans and Sudanese that I met in connection with the study how much money I had. Some could not understand why a person who came from Europe would use the wretched public transport. Some wondered why I did not buy a car. Although I did my best to tell them about my financial reality, some of them interpreted this as I was understating my financial status in order not to help them.

The refugees' stereotypical perception of me was reinforced by the fact that I attended status determination interviews at COR's office and meetings on repatriation. The fact that the refugees witnessed my attendance in these meetings reinforced their belief that I was in a position of power and could help them if I wanted to do so. Although I am not sure how much this affected the reliability of the data I gathered, there is no doubt that the quality of some of the data was affected by the respondents' erroneous beliefs regarding my identity and status.

What is truth? A great deal can be said about the complexity of the concept of truth. One of the things that

I noticed among the refugees in Kassala is that most of them, even the fit, tried to present themselves as being helpless and vulnerable. However, I was unable to determine if they had an interest in creating such a picture. It was not only the refugees who tried to manipulate and use information in pursuit of different ends, the same was true of the staff at COR and UNHCR. When I approached them seeking information, they would try to extract information from me about the intentions of the refugees concerning repatriation. I would tell them that divulging such information would destroy the trust of the informants but they would still insist on that I should share these data with them. I am not sure how much my lack of cooperation affected the reliability of the data elicited through interviews with the staff of these two agencies.

A common problem I had to deal with and found agonizing was lack of punctuality. People never came at the agreed time and place. Most of them showed up several hours later than the agreed time. For a researcher operating under heavy time and financial constraints, this can be a frustrating experience which is sometimes difficult to deal with. Coming from Sweden where people a have strong sense of punctuality made it more difficult for me to cope with the situation. However, I often put up a face and tried to be nice and talk to them in a polite way even though I fumed inside. Some never showed up to meetings. Some of the most useful data were gathered in settings that were unplanned. As a researcher, one is dealing with so many uncertainties and one should be ready and willing to deal with the situation as it arises.

The refugee situation in Sudan at the time the main fieldwork was conducted was highly politicized and polarized. Many political actors create and reproduce difference and polarity, and use every piece of information for their own political end. I find the present development within Eritrean transnational communities problematic since there is a tendency to misconstrue or distort the findings of researchers in order to serve political ends that have nothing to do with the objectives of the studies concerned. The same is true of the information passed on to me by the refugee respondents.

Being an 'insider' there were many things that I saw within the refugee community in Kassala that were very important to my study but which were withheld by the refugees. So unfortunately, there is some information that would have contributed to this thesis but which has been left out for ethical reasons. Although my being an 'insider' was important in accessing certain information which would have been inaccessible to an 'outsider,' the sense of duty I felt towards the community I belonged to was something of a handicap due to the sense of guilt I often felt at not being in a position to help when help was needed. I have tried to walk a tight rope on the one hand, to do justice to the research questions I posed for the thesis, and on the other, to respect the boundaries set by my informants even when I thought that the boundaries were intentionally created to prevent the truth from being known to outsiders.

Some Remarks on Objectivity

Commenting on the draft version of this thesis, Dr. Ali Najib of Uppsala University, raised some critical questions regarding issues of objectivity. He saw a potential risk of a lack of objectivity, which results from a researcher's deep involvement in the situation of the subjects of study. The judgment and objectivity of a sympathetic researcher may be clouded and consequently become biased and less accurate in analysis of the facts. Secondly, the behaviour of respondents may be influenced in pursuit of certain goals due to their awareness of a researcher's empathy towards them. The question to ask is: To what extent has my objectivity been affected by my shared identity, experience, and empathy? I have to admit that my experiences during the early journeys are very similar to the experiences of the majority of the respondents. Undoubtedly, I am sympathetic with their situation. The question is whether this empathy has shadowed my judgment in any way and produced biased results that do not lend reliability to their narratives. Although it is difficult to determine this with absolute certainty, I have done everything within my means to analyze carefully the data obtained from the respondents. As stated above, I was aware from the outset that the situation within which the refugees operated was highly politicized and I expected the views of some of the respondents to be biased. That was the reason I employed different methods of data collection and relied on diverse data sources.

When I first arrived, some of the refugees associated me with aid-giving organizations. This would have definitely influenced their behaviour. In order to avoid this danger, I made sure that the respondents understood that I was an independent researcher who had no connection with governments or aid agencies. After a while, the refugees could see for themselves that I was a student.

The question of validity refers to the extent to which one is asking the right questions to understand the stated objectives of the study, as well as to the appropriateness of the selected method. The degree of validity of the study was maximized by the pilot study conducted prior to the main fieldwork and substantially contributed to the improvement of the research instruments. My familiarity with the culture and way of life of the refugees, as well as my fluency in the languages and symbols of the subjects of my study also improved the validity of my approach.

Given the opposition of the refugees towards the Eritrean government and the ruling party, it is reasonable to expect that their opinions in this regard cannot be taken for granted. That is why I have subjected their answers to precise critique and tried to countercheck them through information received from varied sources, including Amnesty International and Human Rights Watch. The most difficult methodological question this study poses is the extent to which the respondents are telling their story with regard to the factors that determined their decision to stay in Sudan instead of returning to Eritrea in response to the political change that took place in 1993. Which of the many factors identified in the empirical chapters were decisive in determining the respondents' decision? Are some of the country-of-origin related issues reported by the respondents relevant to their decision? Was the decision of the refugees to stay in Sudan influenced by conditions in Sudan, or by the dream of resettlement to the countries of north and Australia? There is no easy answer to these difficult questions.

Analysis

The interviews were conducted in four languages and therefore had to be translated into English. After the transcription of the interviews, the factors that influenced the decision of the refugees were identified; an effort was also made to identify the relationship between the identified factors. Moreover, in order to facilitate these tasks, I drew up an illustrative figure that identified the key factors and the threads that linked them to each other. After performing this, an attempt was made to prepare a thematic scheme that classified the main factors that underpinned the process of decision-making.

The transcription made it possible to identify the recurring themes and the critical issues. Once the data were transcribed, processed and the important issues identified, it was necessary to create categories from the data and to develop structure for the categories. From the classification process, an effort was made to create different thematic subdivisions consisting of different variables (Nieswiadomy,1998).

The data elicited from the interviewees were analyzed manually. Thereafter I organized systematically the field notes derived from informal discussions, focus group discussions and my own observations. Not only did these data supplement the interview material, but I also used them to countercheck the information gathered from the interviews and other data sources. This was extremely useful. Notwithstanding the fact that the data from different sources were collected using different methods of data gathering, their similarities were astonishingly high.

NOTES:

1. There are several reasons for this. According to informal information, the Sudanese government has been at odds with the Eritrean government since 1993. This is because of Sudan's alleged support for the Eritrean Islamic Jihad Movement (EIJM), which engages in sporadic cross-border armed incursions. In retaliation, the Eritrean government has been providing open support to Sudanese opposition groups based in Eritrea, especially the National Democratic Alliance (NDA). Nearly all the Eritrean opposition groups that aim to overthrow the Eritrean government are also based in Sudan, especially in Kassala town.
2. Town in eastern Sudan.
3. Muraba'at, Abukamsa and Souq, October – November, 2002.
4. Khartoum, September, 2002.

4

ERITREA AND SUDAN

This chapter contains two parts. In the first part I shall bring up some background information about Eritrea so as to help us understand why the refugees left their country of origin. It also presents the cultural context of Eritrea regarding religion, ethnicity and gender. In the second part, I shall present an overview of Sudan, the refugees' country of asylum. It also presents some information about the culture and demography of the Kassala district and Kassala town. Moreover, the chapter accounts for socio-cultural similarities that exist between the locals in the study area (Kassala) and some sections of the Eritrean refugee population.

Like most other African countries Eritrea in its present territorial shape was a colonial creation. Prior to the Italian colonial occupation, different parts of the area now known as Eritrea were ruled by successive external powers – including Abyssinian kings, the Fung of Sennar (a Sudanese clan leader), Ottoman Turks and Egyptians

(Habte Selassie, 1980; 1989; Kibreab, 2005). Between 1885 and 1941 Eritrea was an Italian colony (Bondestam, 1989; Kibreab, 2005; Barrera, 1996). When Italy was defeated by the Allied Forces in World War II, Eritrea was placed under British military administration, which lasted until 1952 (Habte Selassie, 1989; Kibreab, 2005; Markakis, 1990). As we shall see below, the future rule of Eritrea was highly contested right from the outset (see Kibreab, 2005). In 1952 the United Nations General Assembly contrary to the expressed wish of the Eritrean people federated Eritrea with Ethiopia under the sovereignty of the Emperor of Ethiopia (see Habte Sellasie, 1989; Bondestam, 1989).

The general expectation among the Eritrean people was that they would be given an opportunity to determine their destiny at the end of the British Military Administration. However, the governments of Great Britain and the United States of America felt that their strategic interests would be served better if Eritrea were annexed to their ally Ethiopia, under the rule of Emperor Haile Selassie. Instead of honouring the terms of the federal arrangement the Ethiopian government systematically undermined the autonomy of the Eritrean government and unleashed violent attacks on Eritrean nationalists and trade unionists and consequently paved the way for the country's absorption into the Empire. When every opportunity of peaceful resistance was exhausted, many Eritreans began fleeing the country to seek protection and to continue the struggle from exile in Sudan and Egypt. The Eritrean Liberation Front was therefore established in 1961 to fight Ethiopian annexation. In 1962 the Ethiopian government, completely disregarding the expressed wishes of the Eritrean people, reduced Eritrea into Ethiopia's fourteenth province. It took thirty years of bloody war to realise independence with the loss of many lives and the massive destruction of property and productive assets. Ethiopian forces were thrown out of the country in May 1991. The country achieved its formal independence after 99.8 percent of the population voted in favour of independence in April 2003.

Eritrea is one of the new independent countries in the world. It is located to the north east of Ethiopia and it borders the Red Sea between Sudan and Djibouti. The coastal plains give way to deserts in the north and west. In the middle, the land rises to the plateau which is a continuation of the Ethiopian highlands in which Asmara, the capital, is situated. Eritrea has a total area of 121,320 km^2, and the capital city is Asmara. The total population is about 4.5 million (www.shabite.com). There are nine linguistic groups in the country. They speak Tigre, Tigrinya, Arabic, Blin, Nara, Saho, Kunama and Afar. The population of the country is 50 percent Christian and 50 percent Moslem. The country has no official languages. However, Arabic, Tigrinya and English are regarded as working languages. The currency of the country is called the Nakfa. Eritrea is a very poor country and the country relies heavily on its diasporic communities for foreign exchange. Not only do the Eritrean diasporic communities send remittances to their families and relatives. Some of them also send money to support the government. These are the ones that support the regime in Eritrea. However, some authors think about them as a group who substantial contributions for post-conflict reconstruction. Agriculture is the main stay of Eritrea's economy. Although by African standards, Eritrea had well-developed industries in the 1930s, 1940s and 1950s (see Trevaskis, 1980), many decades of war and neglect have undermined the manufacturing sector.

Eritrea

Eritrea under Ethiopian Rule

The Italian colonial rulers were defeated by the Allied Forces at the beginning of World War II and as a result Italian occupation of Eritrea ended in 1941. Eritrea was placed under the trusteeship of British military administration until 1952. After the second half of British administration, Eritreans were allowed to form, professional associations and political parties in order to shape their future. This period saw some conflicts between some sections of the country's Christians and Moslems. In the 1940s, the majority of those who were in favour of unification with Ethiopia were Christians and those who were in favour of independence were Moslems. (Ammar, 1992; Interview with one of the pioneers of the Eritrean armed struggle, in Melbourne, Australia, 1999; Ellingson, 1977).

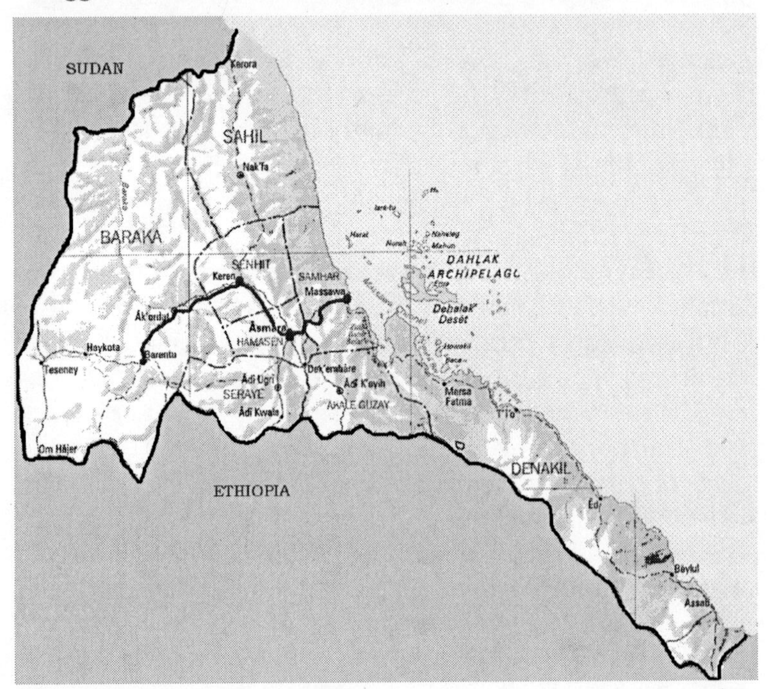

1. Map of Eritrea. Reproduced with the permission of Michael Miller, the owner of the web site www.rightsmaps.com. Date of access: 7 December, 2006.

The British military administration pursued a policy of divide and rule. This was reflected in its plan to dismember Eritrea along religious and ethnic lines against the wishes of the Eritrean people (Kibreab, 2005). If Britain had had its way, Eritrea would have been dismembered in which the Plateau and the coastal plains would have been handed over to Ethiopia and the western lowlands to Anglo-Egyptian Sudan. There is a lot written on this period of Eritrean history and no attempt is made to repeat what has already been dealt with in the literature in great detail (see Markakis, 1990; Habte Selassie, 1989; Kibreab, 2005).

When the four major powers, namely the USA, UK, France and the Soviet Union, failed to reach an agreement on the future status of the former Italian colonies, the cases were referred to the UN General Assembly. In order to determine the wish of the Eritrean people the UN sent a commission of inquiry to look more closely at the situation in 1949. The members of the UN commission could not agree and recommended different options: Independence or federation with Ethiopia. According to General assembly resolution 390-A (V) adopted on 2 December 1950, Eritrea was federated with Ethiopia under the sovereignty of the Emperor. According to the resolution, Eritrea was to have its own government.

Believing that federation under these terms would enable the Eritrean people to lawfully defend their interests, they grudgingly accepted the proposed federation. However, it was hotly debated for two years before the federation was formally in place. In spite of charges of gross irregularities, the federation between Ethiopia and Eritrea became effective on the 15th of September 1952.

British rule came to an end in September 1952 and the Eritrean government was formed with its own executive, legislative and judiciary. Eritrea had its own democratic constitution, which guaranteed the Eritrean people freedom of press, organisation, and movement. None of these rights were known in Ethiopia and therefore the two countries could not coexist very long. Either Ethiopia had to catch up with Eritrea, which was not easy given the autocratic nature of the Emperor's rule, or Eritrea had to regress in order to become part the feudal regime. The Emperor opted for the latter and therefore the Ethiopian government undermined systematically the establishment of the federal Eritrean government. As a result, the Eritrean government was faced with some insurmountable problems from the outset because Ethiopia was determined to annex Eritrea as one of its 14 provinces (Ammar, 1992).

Between 1952 and 1961 the autonomy of Eritrea was eroded systematically by Ethiopia. One by one, the guaranteed rights were violated. In the hope and on the assumption that the United Nations would carry out its responsibility, Eritrea sent special delegations and petitioned the United Nations to stop Ethiopian encroachment, but to no avail. The international blanket of silence over Ethiopia's infringement of Eritrean autonomy and the rights of its people was due to the strategic interests of the big powers, particularly and USA and the UK (Kibreab, 2005a). The USA had an intelligence-gathering base in Asmara at Kagnew Station and had military agreements with Ethiopia.

On 14 November 1962, some members of the Eritrean assembly were ordered at gunpoint by Ethiopian forces to meet and dissolve the federation. The federation of Ethiopia and Eritrea was declared null and void and Eritrea was annexed into Ethiopia. This siege of Eritrea was a UN responsibility but the United Nations avoided intervention, something that is remembered vividly in Eritrea today. Eritrea still seems to hold bitterness against the United Nations and does not have faith in its integrity (Mcspadden, 2002).

Political Resistance

Oppression engenders resistance, and this was what happened after Eritrea's annexation into Ethiopia. The Eritrean people's peaceful resistance produced no results and this gave birth to an armed struggle The Eritrean Liberation front (ELF) was founded in 1960 by Eritrean nationalists who were exiled in Cairo. The founders of the ELF were Moslems but later they were joined by some

Christian compatriots. The armed struggle began in September 1961, in the western parts of Eritrea. Owing to the fact that some prominent Christians were in favour of unity with Ethiopia during the 1940s and the early 1950s, the Moslem leaders of the ELF mistrusted Christians, including those who came to join the Front (Markakis, 1998). This mistrust was based on experiences from 1941 when Eritrean Christians supported unification with Ethiopia and Moslems supported independence. In the mid-1960s, the ELF went through many difficulties occasioned by internal strife. As a result, many of its members abandoned the ELF towards the end of the 1960s. The breakaway groups formed the Eritrean People's Liberation Front (EPLF) in 1972. A bitter and bloody civil war broke out between them (Ammar, 1992).

There were many failed ceasefires and a fiercer civil war broke out in 1978, which led to the ejection of the ELF from Eritrea in July 1981. According to the members of the ELF, the EPLF was supported by the Tigray People's Liberation Front (the present ruling party in Ethiopia, which in 1981 was waging guerrilla warfare to overthrow the Derg regime there). Since then, the EPLF remained as the only front with military presence in Eritrea. The EPLF's victory over the ELF caused bitterness and resentment among the supporters of the latter.

The EPLF continued to fight against Ethiopian occupation and inflicted a defeat on the Ethiopian forces in May 1991. However, the price for independence was high. Many Eritreans were massacred and many became refugees in Sudan and other countries (Kibreab, 1987). In April 1993, over 99.9 percent of the Eritrean people voted in favour of independence in a U.N. supervised referendum and Eritrea became formally independent in May 1993. An exclusionary provisional government was formed 1993 by the EPLF. The Eritrean people expected to see a government of national unity formed by all the fronts that fought for national independence, but the EPLF ignoring the wishes of the Eritrean people formed its own government. Members of all the opposition groups were welcome to return to Eritrea, but only as individuals. Nevertheless, the government was aware of the political and social unrest that could arise if people's interests and needs were ignored (Mcspadden, 2000).

In February 1994 the EPLF held its third organisational congress in Nakfa and adopted a national charter in which it promised to prepare the groups for democratic and multi-party elections. The EPLF also changed its name into the People's Front for Democracy and Justice (PFDJ). On paper, the promises of the national charter were encouraging. Soon after, a constitutional commission was formed and the draft constitution was ratified by the Constituent Assembly in May 1997. However, the government has since then refused to implement it and the country is still ruled by decrees rather than by a constitution. In spite of the rosy promises, Eritrea remains a one party and an oppressive state that tolerates no dissent or any form of opposition (Human Rights Watch, 2005).

Since the ELF's initial areas of operation were the western lowlands in which the large majority of the inhabitants are Moslems, historically the Front was perceived to represent the interest of Moslem Eritreans. In reality, the ELF is a nationalist rather than a sectarian organisation (Ammar, 1992). Its ejection from Eritrea was resented by some Moslem groups who saw the victory of the EPLF as representing the victory of Christians over Moslems. This led to an unprecedented degree of polarisation.

In 1999, the excluded opposition groups met in Khartoum and created an alliance to fight against the regime in Asmara. In 2002 eleven opposition groups met in Addis Ababa under the sponsorship of the Ethiopian government and formed the Eritrean National Alliance (ENA). In 2005 sixteen opposition groups, including the members of the ENA formed the Eritrean Democratic Alliance (EDA) to fight against the government in Asmara (Discussions with Eritrea opposition groups in Diaspora, 2005).

Politicisation of Eritrean Society

During the war of independence, all Eritreans overlooking their religious, ethnic and regional differences fought together against a common enemy. In the process strong bonds of trust were developed. However, in the post-independence period, the exclusion of the deserving political organisations from power sharing is having a very negative impact on the relations between the groups that support different political organisations.

Although the ejection of the ELF by the EPLF in 1981 had engendered a sense of resentment, this would have been overcome had the EPLF opted for a government of national unity. However, since this did not happen, the resentment of some Moslem groups intensified in the post-independence period. For example, the large majority of the Eritrean refugees who have stayed in Sudan are Moslems who had strong links with the ELF.

In May 2002, the UNHCR announced that most of the Eritrean refugees in Sudan would no longer receive automatic refugee status. According to UNHCR the conditions that forced the refugees had ceased due to Eritrea's independence. Hence the UNHCR declared that all those who had fled due to the war of independence and the border war should return to their country, Eritrea, by the end of the year. Those who feared persecution upon return were asked to reapply for a new refugee status on individual basis.

A new Refugee Status Determination (RSD) programme was established to screen the new claims. The RSD was run parallel to the repatriation programme. The RSD began in September 2002 and was supposed to end in December 2002, but the process took much more time than anticipated. The reason was that more refugees wanted to reapply to stay in Sudan than those who wanted to return. Those who were unwilling to return had to pass the new refugee status determination procedure. People hoped that if they passed the RSD, they would be resettled in third countries, and that was one of the major factors that militated against repatriation (interview with UNHCR officer, 2003).

I know from my fieldwork that many refugees who did not pass the RSD test went to register for return migration. I know also from my interviews that many refugees were reluctant to return to Eritrea even before the RSD programme started.

Language, Ethnicity and Holidays

Eritrea is inhabited by about 50 percent Moslems and 50 percent Christians. A small fraction of the population adheres to indigenous religion. Officially there are nine recognised ethnic groups. These are Kunama, Nara, Blin, Tigre, Tigrinya, Rashyda, Saho, Dankel and Hedareb. The Tigre and the Tigrinya are the majority and the rest are minorities. Most Christians live in the highlands while most Moslems live in the lowlands. The dominant groups among the Christians are the Copts whilst the Moslems are predominantly Sunni (Pool, 2001).

The post-independence government has adopted a mother-tongue policy in primary education in which all children study in their mother tongues and after primary school, English becomes the medium of instruction. Arabic and Tigrinya are taught as languages in primary school even in the schools where a mother tongue is used as a medium of instruction. The Eritrean government does not have a policy of an official language. Arabic and Tigrinya are declared as working languages. However, the government's critics argue that actually Tigrinya is the undeclared official language of the country whilst Arabic is relegated to the background. In the interest of equality between Moslems and Christians, in the areas controlled by the EPLF, Wednesday was observed as a weekend whilst Friday and Sunday were normal working week days. After independence, government offices were closed on Saturdays and Sundays whilst Friday was regarded as a working day (interview with staff at the Eritrean council in Khartoum, November 2002).

Violation of Human Rights in Post-independence Eritrea

Eritrea is a one party state. According to Human Rights Watch (2004), no other political party is allowed than the People's Front for Democracy and Justice and no group larger than seven persons is permitted to assemble without government authorization. No national election has ever been held since the country gained its independence. An election was planned for 1997, but it was cancelled, allegedly due to the border war with Ethiopia. The war ended in 2000. In Eritrea, national service is compulsory for all men and women between 18–40 years. The government does not even recognise the rights of conscientious objectors. National service includes six months military training and 12 months participation in national reconstruction. However, since the 1998-2000, national service in Eritrea has been prolonged indefinitely.

As seen earlier the government has refused to implement the constitution that was ratified by the National Assembly and unconstitutional rule has been one of the major causes of problems facing post-independence Eritrea. Had there been a constitution, there would have been no room for arbitrary rule as is the case at present. According to Human Rights Watch's 2004 report:

> "The constitution contains restraints on the arbitrary use of power. It provides for writs of habeas corpus, the right of prisoners to have the validity of their detention decided by a court, and fair and public trials. The constitution protects freedom of the press, speech, and peaceful assembly. It authorizes the right to form political organization. It allows every Eritrean to practice any religion." (ibid. p. 3)

The report documents a series of violations of human rights

– e.g. suppression of minority religious groups, forcible recruitment into national service, arbitrary detention under appalling prison conditions and torture.

The government uses the border war as an excuse for postponing the implementation of the constitution and postponement of democratic elections. Amnesty International's report (2004) also shows that although the constitution guarantees the right to freedom of religion, Jehovah's witnesses in Eritrea (who number about 1,600) are suppressed and detained arbitrarily for refusing to join the national service, which is against their faith. They are also stripped of their citizenship rights. The government even closes down other minority churches, such as the Pentecostal and evangelical churches. These churches were asked to register with the department of religious affairs. However, although they applied, not one has been given a permit so far. In 2003 over 330 members of these faiths were arrested and held incommunicado detention without being charged. Many have been tortured or ill-treated in the attempt to force them to abandon their faiths. Former detainees at the Dahlak Island in the Red Sea where up to 8,000 inmates are held incommunicado in detention have recently testified that whoever was caught praying was tortured. Religious books and cassettes were also burned.

Gender Relations in Eritrean Society

In Eritrean society, gender relations are deeply rooted in the realms of culture, religion and economics. Social relations are strictly gendered. All the ethno-linguistic groups have their own gender ideologies that determine the rules governing division of labour, ownership of property, including land, and decision-making within households and communities. Women alone bear responsibility for all types of reproductive activities ranging from childcare, food processing, cooking, cleaning, laundering, socialisation of children, nurturing of men and children, caring for the sick family members and neighbours, etc. Depending on the specific culture of the various ethno-linguistic groups, women are also responsible for food crop production (Kibreab, 1992).

Men are invariably seen as being the breadwinners of their families even in situations in which a female member may be the only or the main contributor to her family's income (Kibreab, 2003; Gruber, 1999). Eritrean culture prescribes what women and men are expected to do and what they are not supposed to do. For instance, the traditional Eritrean social norms do not allow female participation in politics or in other activities that take place in the public sphere. This restriction is due to the predominant assumption that the activities of women should be confined to the private sphere and everything that takes place within the public sphere is supposed be the exclusive domain of men.

Men and women are socialised from early childhood into pre-determined relations of inequality and over time the norms and values that are systematically inculcated to create differences and inequalities between the two sexes become normalised. As a result, women are subordinate to men, who exercise power and control over women. In this respect Eritrea has been, and still largely is, a patriarchal society.

Religion and Gender Roles

In Eritrea, the state is secular and the civil, criminal and commercial laws of the country are secular and therefore are not clearly influenced by religion. However, because Eritrean society is deeply religious and the "prescriptions" of Islam and Christianity have substantial influence in the way they conduct their lives, especially in areas pertaining to family law. Both Christianity and Islam have their own specific rules and norms, which are supposed to regulate the relationship between the two sexes. Moslems and Christians are expected to be bound by these rules. In Islam, female and male siblings do not have equal rights to inherit property of their parents, and married women are only entitled to one-eighth of her husbands' property. The property may in actual fact belong to both husband and wife but is referred to as being the property of the husband. Moreover, in the current interpretation of Islam women are not allowed to work outside the family home without the permission of their husbands Hassanen, 1997, 2002).

It is important to note, however, that the followers of the two religions may not comply with the scripts of the two Holy Books. For example, In Islam alcohol is *haram* (forbidden), but there are many Eritrean Moslem men and women who consume alcohol. In Christianity, polygamy is strictly prohibited, but it is common knowledge that there are many men who have mistresses. However, this does not mean that all polygamous Moslems do have mistresses.

In Eritrea, there are three types of laws: Secular civil law, customary law and Shari'a law. Customary and Shari'a laws are applied in family matters, such as marriage, child custody, inheritance and divorce. Although customary and the Shari'a laws originate from different sources, they both treat women as being subordinate to men. Both laws affect Eritrean women more or less in the same way. For instance, in Islam in case of divorces, the husband has the exclusive right of custody of children. However, children under seven years can stay with their mother until they are older and then they are handed over to their father. Polygamy is widely practised in Islam. Islam allows men to have up to four wives provided they can provide for them all (ibid.).

Although the rate of illiteracy is generally high in Eritrean society, it is much higher among women than men. This is because of the deeply rooted societal belief that investment in boys' education is regarded as being beneficial to investors – families – whilst investment in women's education is considered as bad investment because it is the future husband and his families that reap the benefits, not those who met the costs of investment (ibid.).

Patriarchy, Ethnicity, Poverty and Change of Gendered Tasks

Eritrean society is patriarchal. Men are supposed to be independent breadwinners, emotionally restrained, sexually dominant and authoritarian whilst women on the other hand, are expected to be dependent, relegated to domestic work and devoted to their husbands and children. However, Eritrean society consists of different ethnic and religious groups and it is not possible to generalise the manner in which women are treated. Each ethno-linguistic group has its own codes of conduct regarding the relations between the two sexes (Kibreab, 2003).

Eritrean society is undergoing some fundamental transformations in regard to gender relations. There are many families that do not observe the gender-based divisions of labour or the roles assigned to men and women. The thirty years war has brought about some fundamental changes in the manner in which men and women relate with each other. For example, in the EPLF, women represented over 30 percent of the fighting force and participated in all activities, including in the frontlines. Although there have been some indications that suggest that some of the gains women made during the war are being lost in the post-independence period. There is no going back to the past. Female participation in the labour market has, for example, become a norm in post-independence Eritrea, especially in the urban areas. (Pateman and Roy, 2003; Kibreab, 2003; Connell, 1998).

Technological change has also lessened the burden of women's reproductive work. For example, the backbreaking task of grinding of grain has been eased by the introduction of grinding mills even in rural Eritrea. The introduction of piped water has also eased the burden of water fetching, a task invariably performed by women. In his study of the refugee settlements in eastern Sudan, Kibreab found that shortage of firewood led to the emergence of a firewood market (Kibreab, 1996c). This indirectly lessened women's burden of domestic activities. The same trend is visible in many parts of Eritrea. Forced migration has also brought some changes in gender-based divisions of labour among Eritrean refugees. Tasks that were previously regarded as the exclusive domain of women have ceased being so (Kibreab, 1995; Moussa, 1993).

Historical Links between Sudan and Eritrea

The historical links between Sudan and Eritrea are important for this study. Western Eritrea and Eastern Sudan have very similar land formations, climatic conditions and the inhabitants of both areas share similar histories, cultures, norms and values. For instance, the Red sea hills in Sudan are the continuation of the hills that extend to the Eritrean Sahil region. There are also similarities between the populations inhabiting these areas. The earliest inhabitants of Eritrea are believed to be the Nilotic dark skinned people who moved from their original habitat in the south-eastern region of present Sudan into the Eritrean lowlands of the Gash and Setit valleys (Goitom, 1980). The Beja are the other group who settled in Eritrea. They are the descendants of the nomadic Hamitic clans from the deserts of northern Sudan that occupied the western lowlands along the Baraka River and the northern highlands. The present lineages of Beja are the Hadendawa and the Beni Amer (ibid.).

These clans occupy the area on both sides of the borders. In Sudan, the Beja clans, mostly the Beni-Amer and the Hadendawa, are concentrated in the towns of Port Sudan and Kassala in Eastern Sudan. The dominant clan in western Eritrea is the Tigre-speaking Beni Amer who are Moslems. The border Eritrean and Sudanese communities, therefore, share common history and ancestry (Pool, 1992).

The relationship between the border communities is marked by interdependence and intensive economic and social interactions. These economic relations also continued during the thirty years' war. Most of the cross-border trade takes place illegally. The Sudanese and Eritrean markets complement each other and the border communities derive a substantial proportion of their incomes from such activities. Even before the border war, labour migration from Eritrea to eastern Sudan was common (Ammar, 1992).

"Most of the labour migrants were unskilled and they worked in the rain-fed mechanised agricultural schemes and in the irrigation schemes in eastern Sudan. Although the large majority were seasonal migrants, some of them ended up staying in Sudan. Most of those who stayed in Sudan, acquired Sudanese nationality.[1] These historical links are critical to the way Eritrean refugees adapt to life in Sudan. As we shall see later, common clan membership is an important factor in the process of integration of Eritrean migrants and refugees in eastern Sudan." (Kuhlman, 1992)

The majority of Eritrean refugees in Sudan are Moslems, many with a previous relationship to the Eritrean Liberation Front (ELF), which opposed the EPLF during the liberation struggle (Mcspadden, 2000, p.73).

Sudan

In terms of territory Sudan is the largest country in Africa. It covers an area of 2.5 million square kilometres and has a population of about 25 million (Länder i fickformat, 2002). Even though Sudan is a poor country, it is endowed with considerably more natural resources than Eritrea (Kibreab, 1983). The country is divided into five regions; the southern, central, northern, western and eastern regions. When people talk about Sudan, they generally refer to the northern and southern regions of the country. This is due to the wide religious, racial and cultural differences that exist between the north and the south.

Arabic speaking tribes, who are descendents of the indigenous peoples and Arab migrants who came to the region over the past 1300 years bringing with them the Islamic culture from the Middle East, inhabit the northern regions of the country. The southerners are mainly Christians, but there are also animists. Missionaries spread Christianity in this area in the early 20th century (Edward, 2001). The religious and ethnic differences in Sudan between the southern and northern parts of the country have given rise to a serious political conflict that has lasted several decades. Due to the protracted conflict between the South and the North, many southerners fled their homes to neighbouring countries to seek international protection (ibid). Sudan is a major refugee hosting country but at the same time a major refugee producing country, most recently seen in the massive refugee emergency in the Darfur region of western Sudan.

The ethnic groups that are present in both Sudan and Eritrea have the opportunity to be citizens of both countries. This may be somewhat confusing to researchers who are interested in social identity and mobility. The Beja are found both in Eritrea and Sudan. Many people in Kassala belong to the Beja and they can live as Sudanese nationals or as Eritreans. The phenomenon of dual nationality is not limited to the Beja (Beni Amer). There are also other Eritrean clans, such as the Moslem Blin, the Maria and the Bet Juk that also have access to dual nationality. The possession of Sudanese nationality enables the refugees to circumvent the restrictions imposed on refugees with regard to freedom of movement, residence, employment, and access to other rights. According to the Regulation of Asylum Act, 1974, refugees are required to live in government-designated areas and are not allowed to work in the urban centres unless they have a special permission to do so (Kibreab, 1996, p. 19).

The Kassala region encompasses the administrative units of Kassala town and Kassala rural district. Kassala has a semi-arid tropical climate. The rain season lasts about three months per year from June to September. The rain falls in heavy downpours that mostly follow heavy dust storms. The average rainfall is about 286 mm, but this varies from year to year (Kuhlman, 1988). The Gash is the main river in Kassala. The river is seasonal. In good years, the water in the Gash River is used for irrigation. The dominant soil in Kassala is heavy clay which by Sudanese standards is regarded as fertile. However, it is difficult to break without tractor-drawn ploughs (Kibreab, 1987).

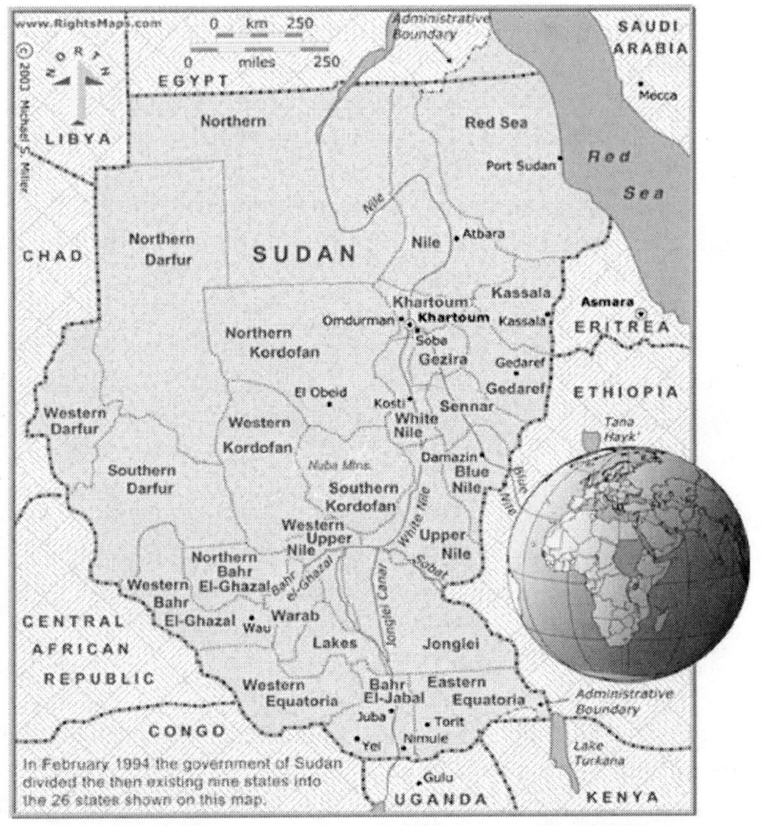

2. Map of Sudan, Reproduced with the permission of Michael Miller, the owner of the Website www.rightsmap.com. Date of access: 7 December, 2006.

Different ethnic groups inhabit the Kassala region. There are some researchers who claim that the majority of the inhabitants are Eritreans who belong to the border areas (Baume, 2000). The reliability of this assertion is, however, questionable. The native inhabitants are the Beja settled both in Eritrea and Sudan. The Beja of Eritrea are settled in the western part of the country.

As Kuhlman (1988) notes, there have been cross border movements between Eritrea and Sudan throughout the twentieth century. Some of the Eritreans worked in Kassala town while others found their livelihood as agricultural labourers in the villages in the Gash delta. In the beginning of 1940, Baria and Baza tribesmen from Western Eritrea started to arrive as seasonal labourers. Later, some Moslems who protested against the Ethiopian regime arrived in Sudan as refugees. They were welcomed as brothers and later they became leaders of the Eritrean Liberation Movement (Kuhlman, 1992, p. 48). Before the arrival of refugees, the cross border migration between the two countries was almost entirely dominated by livelihood motives.

The flow of Eritrean refugees to Kassala started in 1967 (Karadawi, 1999; Kibreab 1987). From that time onwards until the country's independence, refugees arrived in small groups. The large majority were concentrated to the Kassala region. Some were settled in refugee camps. According to the head of the Red Cross in Kassala, the majority of Kassala´s inhabitants are refugees from Eritrea and mostly from the Tigre speaking Beni Amer. Other Eritrean nationals who settled in Kassala besides the Tigre-speaking groups are the Moslem Blin who are bilingual (Tigre and Blin), Nara, Kunama, Saho and Jeberti. There are also a few Christians. As will be seen later, these differences in ethnicity and religion are critical in the process of integration of the refugees into Sudanese societies. For instance, the Sudanese Beni Amer in Sudan welcomed Eritrean Beni Amer when they came as refugees. Some of the refugees were even given land to cultivate (Kok, 1989). Therefore, Eritreans who had members of their clans in Sudan had no problem of being integrated.

According to Kuhlman (1992) the UNHCR in Sudan and the Sudanese government discouraged refugees to establish themselves permanently in urban areas, among others in Kassala town. The policy was to resettle refugees in camps or organized settlements further inland. It was seen as more convenient to relocate the refugees in organised resettlement with international assistance. Additionally, it is believed that if refugees are allowed to settle among the local population, this will represent a burden to the local economy. The other motive is that in an organised settlement, the refugees would be given land, which is abundant in Sudan and would gradually become economically self-sufficient (Commissioner for Refugees, personal interview 2002). The policies of the Sudanese government and the UNHCR have not been entirely successful. This is partly because of lack of resources and the refugees' reluctance to be settled in rural areas. Those who are self-settled receive no assistance because according to the policy of the government; they are required to settle in camps or organised settlements (Kibreab, 1994; Bascom, 1998). For those who do not share common identity with the local population, settlement in urban areas is risky.

Not only is settlement in urban areas illegal but also it is difficult because the refugees in such areas receive no assistance and finding a job is difficult. The cost of living in cities, except for the merchants and for those who live on remittances, is high. Refugees in urban areas are also subjected to continuous harassment and round-ups (Kibreab, 1996c; Karadawi, 1978, 1999). There are several reasons why refugees prefer to live in urban areas rather than in camps (ibid.). The main motive for living in an urban area is the freedom concerning earnings, access to education and health care facilities. The availability of these services is the main pull factor (Kuhlman, 1991).

Some find it difficult to live in refugee camps (rural areas) because of their background, having lived in urban areas before they became refugees. For others, their children's education may be decisive because in refugee camps, there are only primary schools that are poorly staffed and equipped as such. The families whose children are in post-primary education often move to cities where their children can continue pursuing secondary and higher education. Those refugees who can easily be integrated into the local communities can participate in the labour market based on equality with Sudanese citizens and this is a highly attractive option to the refugees. It is also important to note that there are many refugee families who move to Kassala town from the neighbouring refugee camps of Wedsherifey and Shegerab who continue to collect rations notwithstanding the fact that they live in the city. The refugees jealously guard this information and they are therefore reluctant to share it with outsiders, including researchers. However, the fact that this takes palace is common knowledge among the refugee communities in Kassala town and the refugee camps.

The question to ask is: How do urban refugees survive without assistance from NGOs, host governments and UNHCR? According to Kuhlman (1994), Horst (2003a) and Kibreab (1996b), refugees who live in urban areas in Africa survive to quite a large extent through remittances sent by relatives who migrated to Western countries or to the oil rich countries of the Gulf, but mainly by participating legally or illegally in the labour market of host countries.

Islamic, Customary and Secular Laws

In Sudan there are three kinds of laws. These are the Islamic Shari'a law, customary law and secular laws. Even though the government of Sudan has tried to implement the Islamic law, this attempt was not successful due to the presence of different religious groups in the country. For instance, the southern Sudanese opposed the introduction of Shari's law because the majority of these people are Christian or are animists. In some parts of the country, customary law exists side by side with the secular and Shari'a laws. Of these different legal systems, customary laws and Shari'a law carry greater weight with regard to family issues, property ownership, marriage, divorce, child maintenance and custody. The Shari'a and secular laws are written laws while the customary law is not codified (Edward, 2001) The former two legal systems apply more or less to all ethnic groups, especially in northern Sudan, whilst customary laws vary from one clan to another. Customary law and Shari'a law are applied by local courts that are headed by judges or sheikhs. Men hold these positions (ibid.). In some places, the interpretation of Islamic laws is underpinned by customary laws, and this reinforces the patriarchal values and norms. For instance, in cases of divorce, child custody, marriage and other family matters, courts may rule in accordance with Shari'a and customary law.[2]

As noted earlier, Moslems and Christians each represent 50 percent of the population in Eritrea. In eastern Sudan where the majority of the refugees live, and where the study area lies, the dominant religion is Islam. Thus, Islam or the Shari'a inspires people's values and norms. The majority of women in Kassala dress according to Islamic dress code. After the introduction of Shari'a law in the country that began in 1983, adherence to Shari'a became intensified and strengthened through the campaigns of the National Islamic Front in June 1989. Most aspects of social life then became re-islamicised in northern Sudan.

This process of re-islamisation has had a tremendous impact on gender relations in the country, including among Eritrean refugees. Women face serious difficulties because according to the norms and values prescribed by Shari'a law, women are considered to be inferior to men. In view of the fact that the grip of Islam varies between different places, the way women are treated at the hands of men also varies between places and communities. In some of the communities in which collectivist rather than individualist decisions are the norm, women are unable to exercise autonomy (Hassanen, 1996).

It is interesting, therefore, to ask how Eritrean women deal with laws such as the Shari'a that do not recognise individual choice and decision-making. Eritrean Moslem refugees find it relatively easy to adapt to Sudanese society but the Christian refugees are faced with intractable and complicated problems because of their cultural values and norms, which make their absorption into Sudanese communities difficult.

The Town of Kassala

Kassala town lies very close to the Eritrean border. It is the seat of regional government and is thus the capital of the Eastern Region. Kassala town has approximately 150,000 officially registered inhabitants (Baume, 2000).

Kassala is a town of merchants, small craftsmen, civil servants, with no particular manufacturing industry (Kuhlman, 1988, 1992; Wijbrandi, 1986). Kassala has eight health centres and one hospital. In addition, most of the population has access to free health service (medical examinations) but they have to pay for drugs. About 80 percent of the children in Kassala go to school. There is one university known as the University of the East. The university was opened as part of the military government's drive to set up at least one university in every region. Most of the new universities are understaffed and poorly equipped (interview with one of the COR staff).

Kassala town is a market centre in a predominantly agricultural region. The central area of the town is filled with shops, hotels, restaurants, banks, insurance services, etc. In this area there are different markets, namely Souq Kassala and Souq Kudar (vegetable market). These are the biggest markets in the town. There are other market places as well but they are smaller markets for petty trade. Most of the restaurants and cafés lie in this area along with a variety of commercial activities such as goldsmith and clothing shops. The official public sector includes the police force, the hospital, the university of Kassala and public schools. Most of these places lie outside the centre with the exception of the hospital.

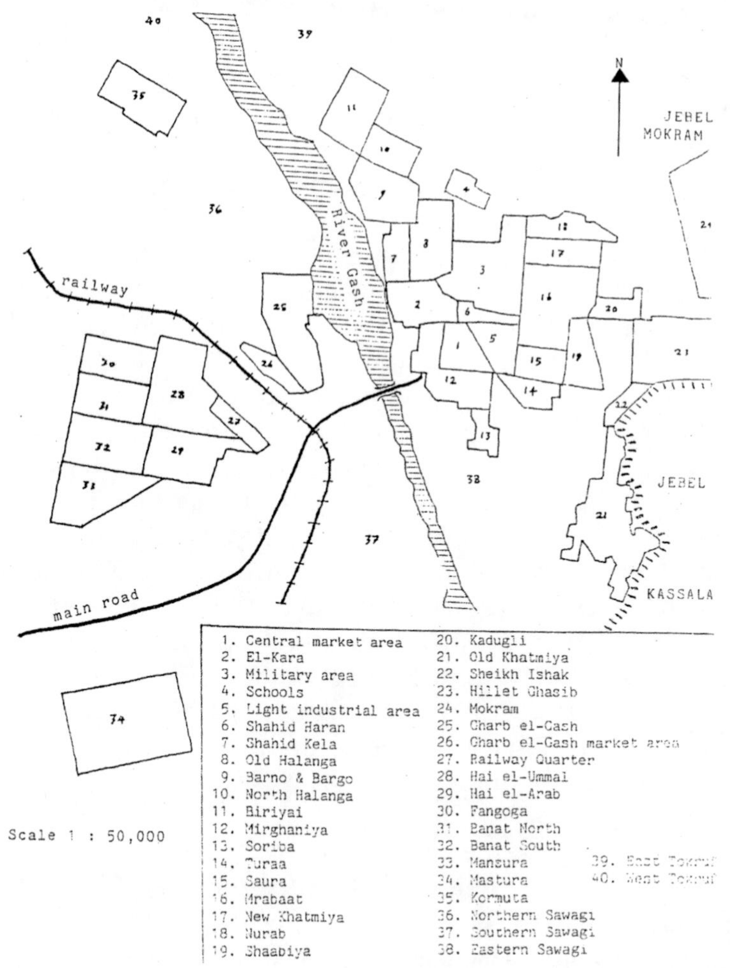

Map 3. Kassala. Copied from Omyma Sead's masters thesis, University of Gedarf, 1988

The town is inhabited by different ethnic groups, such as the Hadendowa, Beni Amer, and Arabs from the Nile valley, Hausa of West African origin, and Rashaida nomads, Yemeni and Indian merchants and Eritreans belonging to a variety of ethnic groups. The residents of Kassala are mostly devout Moslems. About 30 percent of them are members of the Ansar El Sunna (Hassanen, 1996). These people are orthodox and more devout Moslems who are opposed to modernity. They are different from the other sects such as the Khatmiyya and Aquan el-Moslemeen (Moslem Brotherhood). The Moslem Brothers are more modern in their way of living, working and interacting with others.

The majority of the Eritrean refugee camps and settlements are located in the Kassala region. There is a continuous flow of people and goods between the town and the refugee camps and settlements in the region. Many of the refugees in the town were previous residents of the camps or settlements. Those refugees whose children need to attend secondary school tend to relocate to Kassala town from the refugee camps and settlements, provided that they are able to find work and pay rents. Many also move on to other places, including the capital city Khartoum, in search of employment opportunities.

In comparison to the refugees who live in Khartoum, the refugees in Kassala are closer to their own country of origin and they are more integrated into the host society than those who are in Khartoum and camps and settlements. Previous studies on refugees in Kassala show that they are able to visit Eritrea at any time and therefore do not suffer from homesickness (Kok, 1989). Even though the policy of the Sudanese government prohibits refugees to stay in towns, many of them do so. Different researchers have noted (Kuhlman, 1992; Kibreab, 1996; Ibrahim, 2003) that the majority of all refugees in the Eastern region of Sudan are self-settled. Since there are no accurate statistics on Eritrean refugees in Sudan, these are based on estimation rather than on conclusive evidence.

NOTES:

1. These groups are predominantly males. There were women too but they were fewer in numbers.
2. A practicing lawyer: personal interview 2002, Khartoum.

PART 2

FACTORS THAT AFFECT DECISION-MAKING REGARDING RETURN MIGRATION

I have chosen to present my analysis of the empirical data in three separate chapters. Since the factors that affect the refugees' decision-making regarding return migration are multifaceted and interwoven there is a need to structure the presentation so that the complexity of motives is not lost in a simplified catalogue of factors.

The factors that affect the refugees' decision-making are composed of different aspects that intersect each other. For instance, refugees that gave a political reason for not returning also gave social and economic motives for choosing not to return to Eritrea. Two situations in the responses of the refugees show this overlap of factors. As we shall see below the ELF is a political faction, but for some of the respondents it also serves as a substitute for family relations. This indicates that a reason, which can be defined as political on one occasion, can at some other time be social, and vice versa. The need to recognise the complexity of the reasons given by respondents is one of the contributions that this study makes. Another contribution is that of recognising the importance of political factors for people's willingness to return or not. Social and economic issues have dominated earlier studies on return migration. Although, these are important factors it is my contention that the political issues are equally, or even more important.

In the following chapters (5-7) a presentation and discussion of results is given. These three chapters are about factors that influence the decision of the refugees concerning return migration. The factors discussed are: (i) Political conditions in Eritrea, namely country-of-origin related factors; (ii) Socio-economic conditions in Sudan, that is host-country related factors; and (iii) Resettlement opportunities in northern countries, that is migration-to-third-country related factors. After presenting the data, each chapter will conclude by making some remarks on the addressed issue. Chapter eight is a summary of the factors that influence the decision making of the study group and is divided into three sub sections. The first section is about general aspects regarding the decision making factors, the second part is about specific ones and the last one is about the issues that can be studied further.

5

COUNTRY-OF-ORIGIN-RELATED FACTORS

The aim of this part of the chapter is to identify the country-of-origin related factors that influence people's decision making regarding return migration. The political issues addressed in this chapter have not been fully accounted for in previous studies. The data upon which the chapter is based are derived from interviews with key informants, participant observation in the UNHCR information campaigns, discussions with self-settled refugees in Kassala and with refugees placed in the refugee camps. This information was collected in three periods of fieldwork in 2001, 2002 and 2003.

Eritrea was liberated from Ethiopian occupation on 24 May, 1991. The country became formally recognised as an independent state after 99.9 percent of the Eritrean people, including the refugees in Sudan, voted in favour of independence in a United Nations supervised referendum. Since the country's independence, thousands of Eritrean refugees have returned to Eritrea in response to the political changes that have taken place (see Kibreab, 2002; Bascom, 2005). There are still thousands of refugees who have not returned to Eritrea in spite of the elimination of the original factors that caused displacement. There are also thousands who have fled from the independent state, mainly to escape the open-ended national service and from the regime's violation of human rights. Those who abscond or run away from the service are subjected to severe punishments and the best way to avoid being detected is by fleeing to the neighbouring countries, especially Sudan.

The failure of the old caseload refugees to return home has puzzled many observers. This is because when they fled their country, the latter was occupied by Ethiopia. The Eritrean people under the leadership of the EPLF threw out the Ethiopians from the country in May 1991. Other observers expected all the refugees to return home in response to the political changes that had taken place in their country of origin (Kibreab, 1996). However, the government of independent Eritrea did not encourage all the refugees to return. Return is not a mechanistic reaction that occurs in response to political changes. It is far more complicated than that. Many Eritreans have stayed more than thirty years in exile in Sudan. Some were even born in exile without ever having visited Eritrea. After the devastation of Eritrea's economy and infrastructure during the long war of independence and the exclusion of the political organisations, many of the remaining refugees in Sudan who sympathise with the ELF don't regard repatriation as an option that appeals to them. These are some of the factors that have influenced the decision making of the refugees regarding return to Eritrea. From the end of the 1990s, it also became clear to the international community that the Eritrean government was violating human rights (Amnesty International, 2004, 2005) and this serves as a disincentive to some of the refugees.

This is exacerbated by the arrival of new asylum seekers from Eritrea to Sudan. The arrival of new refugees from independent Eritrea indicated to those who had not returned that things were not as they should be in their country of origin. If Eritrea were safe, they reasoned, there would be no need to flee. The new arrivals brought negative news about the country of origin. Refugees are people who flee because of fear of being persecuted or when the bond of trust with their government breaks down (Ghanem, 2003, p. 21). Unless they are sure that the situation has completely stabilised, they are reluctant to return. However, opposing views exist on this question. Most of the reported reasons that account for the refugees' unwillingness to return are associated with the socio-political and economic conditions in the home country. Contrary to these reports a majority of the respondents in this study mention the disregard for democratic rights in Eritrea as the prime reason for their unwillingness to return there. However, it is very difficult to ascertain from the respondents' statements whether their fear is based on real threats to their democratic rights or not. My uncertainty is founded on the fact that while they are expressing these fears other fellow refugees are returning to Eritrea with the repatriation programme. What do these respondents mean by violation of democratic rights? Do they equate the exclusion of ELF from the government with political violation?

The Exclusion of ELF from Power Sharing

During the war of independence, the Eritrean people fought together setting aside their ethnic, religious and regional differences. After the enemy was thrown out of the country, all Eritreans expected and hoped that a government of national unity would be created as a result of agreement between the EPLF and the other organisations that played vital role during the early years of the war of independence. To the disappointment of many Eritreans, the EPLF opted for an exclusionist strategy. The organisations that had legitimate claim on power sharing were left out and could not even return to the country that they had not seen for over thirty years. Of the 24 key respondents, twenty mentioned that the exclusion of the ELF from power was one of many factors that influenced their decision not to return to Eritrea. Even though many previous studies have focused on the political issues among refugees and sometimes among emigrants, few researchers (Kibreab, 1996, Mcspadden, 2000) have seen this issue as an important consideration for decisions people make on whether to return to Eritrea or not.

One of the respondents, for example, said:

> "The majority of the Eritrean refugees in Sudan being Moslems, they never identified themselves with the political system in Eritrea or the EPLF. The Eritrean government is aware of this fact and that is the reason why it ignores us as it does to all Eritreans. Those Eritreans who kept silent about the government's wrong-doings are now raising their voice against the government when they are affected." [1]

Since its ejection from Eritrea in mid-1981, the ELF was split into different divisions. Several of these divisions have followers among the refugees in Sudan. Data obtained from informants among the refugees still in Kassala show that a large majority of those who have not returned are supporters of the excluded political organisations (see also Mcspadden, 1999). This seems to be one of the most important factors that have shaped the attitude of the refugees towards the Eritrean government. In some respondents' view, the reason why the political organisations that they support and sympathise with are excluded is that the political system in Eritrea is undemocratic, which forbids all forms of political organisations except the ruling party.2 Had there been a democratic system, these respondents argue, the ELF and the other factions in exile would have been given an opportunity to compete in a democratic electoral system.3 Some respondents also argue that because the government is undemocratic, it does not recognise the democratic rights of its citizens. Most of the respondents therefore reported that they were unwilling to return to a country where there is lack of freedom and poor governance.

One of the problems that a researcher who relies on data drawn from respondents encounters in a highly politicised environment is the inability to distinguish between genuine and fabricated answers. The Eritrean government's decision in favour of an exclusionist approach of ruling the country has antagonised many of its citizens, particularly those who aspired to share power. The Eritrean government lacks any respect for basic human rights of citizens (Human Rights Watch, 2005; Amnesty International 2002, 2004, 2005). However, it is difficult to determine the extent to which the 'lack of democratic rights' argument influenced the decision of the refugees to stay put in Sudan. Yet the respondents' claim of political alienation cannot be completely ignored. The accuracy of their information or the objectivity of the positions the emigrants have taken about the Eritrean government may seem questionable to some people. Genuine or not, the emigrants' positions indicate that the government has left itself open to negative perceptions by some emigrant communities.

Except for those who are able to show well-founded fear of persecution, their refugee status was to come to an end. As a result, they were expected to return home in response to the political changes that took place in 1991. Kibreab's study in 1994 showed that refugees in Kassala were ready and willing to return. However, my study shows the opposite.

The thirty years war for independence was dominated by the two rival political organisations, the ELF and the EPLF. Although both the ELF and the EPLF fought for national independence, owing to their particular histories, each had followers in different parts of the Eritrea. Until the first half of the 1970s, the ELF was dominant in the Western lowlands of Eritrea whilst the EPLF was dominant in the northern parts of the country. Although the reality on the ground defies any neat classification, there is a general perception or misconception among Eritreans that the ELF represents the interest of Eritrean Moslem lowlanders. For the same reason, the EPLF is assumed to represent the interests of Tigrinya-speaking highlanders. This suggestion gives a simplified demographic explanation. For example, the social composition of the ELF was transformed after the influx of Christian highlanders in the mid-1970s. Similarly, the populations of the Red Sea province, including the Saho, as well as the Habab and the Maria are represented in the EPLF. Regardless of the reality on the ground, such misconceptions were widespread among refugees covered in this study.4

The ELF is an important part of the Eritrean history of armed struggle. The exclusion of the ELF from the Eritrean political scene created bitterness among ELF members. It was a severe blow to those sympathising with the organisation and this definitely had an impact on the decision of those still in Sudan not to return to Eritrea.

As the government in power is assumed to represent the Tigrinya-speaking highlanders, the Kassala refugees feel that their interests are neglected by the exclusion of the ELF from the national process. For they believe that it is only the ELF that can represent their interests. Both the ELF and EPLF have since the 1970s been more than just regionalist political movement. Thus, some Moslem and other groups feel unjustifiably excluded from the social and political life of their country. It seems that the present Eritrean government underestimated the influence of the ELF among the Eritrean people and also among the refugees in Sudan. If the return of the refugees had been considered a vital issue of national interest by the Eritrean government, probably it would have adopted a different attitude towards the ELF if not for anything else but to secure the return of its citizens. The fact that the refugees are upset by the Eritrean government's decision to exclude the ELF from power may be demonstrated by the statement of a family interviewed by me in Kassala town in 2002. When asked to state the reason why they stayed in Sudan instead of returning to Eritrea one interviewee said:

> "The decision of the Eritrean government to exclude the ELF from power is a clear indication that the situation is unlikely to improve in the future. To me, the government's exclusionary approach indicates that it is unable to bring about justice. If the EPLF cared about justice, it would not have ejected the ELF from Eritrea in 1981."[5]

Some of the respondents argue that Eritrea is a heterogonous society. The corollary of this is that their interests are to some extent different and therefore cannot be represented by a single political party. Inasmuch as Eritrean society is diverse, there should be a pluralistic political system that gives expression to different voices and visions. A one party state cannot represent the multiple interests of the society. This is exacerbated by the fact that many refugees view the government in place as being biased against Moslem interests. All the Moslem respondents expressed this view, even those who used to belong to the ruling political party in Eritrea.

> "The government in Eritrea is following the old slogan of EPLF 'We and our goals'. This is what Isayas, the head of the ruling party in Eritrea, used to say to his followers. After independence, he made sure to use the same tune without stating it openly. Until 1998 he confused the people of Eritrea by making changes that do not suit Eritrean people but only him and his followers and by blaming Moslem Eritrean refugees for being terrorists and enemies of Eritrea. He says this to stay in power. Since 1998, he has been using the border war as a means of scaring the Eritrean people. He uses these excuses to postpone democratic elections and to prolong his dictatorial rule of the country."[6]

In this respondent's view, in a multi-cultural society such as Eritrea, democratic pluralism is essential. This is because of the opportunity it offers its citizens to choose among different candidates that they may believe best represent their interests. If there is no choice of candidates, the persons who win the elections may not be considered as having legitimate claim to power. This may engender resistance or opposition against such a regime by those who feel excluded. The data elicited from respondents indicate that they had high hopes that independent Eritrea would be a place of respect, solidarity and participatory democracy. Unfortunately, their expectations are frustrated and one way the refugees express their resistance appears to be by staying away from the domain of the government they consider undemocratic. Had they been given an opportunity to vote and to stand for elections, they would have felt included. This would have reconnected them with their home country and increased their commitment. This signifies that for these informants and others like them the feeling of home is associated with the acceptance or rejection of the ELF. If the ELF is accepted in the Eritrean political scene, then Eritrea could once again become home. However, if the ELF is rejected by the Eritrean government, then Eritrea will not be regarded as home.

Unfulfilled Dreams

Prior to and in connection with their displacement, the large majority of the refugees had suffered at the hands of the Ethiopian government. "Had it not been for our commitment to the war of independence and to the political organisations that were determined to bring about change," a group of refugees argue, "we would not have been subjected to abuse."[7] Therefore, many of the key respondents regarded themselves as being an integral part of the revolution. In other words, they looked at themselves as being freedom fighters. This perception undoubtedly engendered a powerful feeling and expectation that they would be received as heroes in post-independence Eritrea. When none of this happened, they felt seized by disappointment and resentment. One of the respondents, for example, said:

> "My dream was to live in a free and liberated Eritrea, the country of my ancestors. I did not expect my freedom to be limited by restrictions imposed by the Tigrinya speakers. The government has broken the Eritrean people's heart by doing what Ethiopia did against the Moslem communities after the federation. What is happening in present day Eritrea is not different. Had we known the outcome, we would not have participated in the struggle. The ELF will continue its fight until the dream of the Eritrean people is realised."[8]

The data collected from key respondents in Kassala indicate that the large majority of the refugees are either active members or sympathisers of the ELF. However, none of the respondents held at the time a leadership position in the organisation. The general impression I got from the various interviews I conducted among the refugee communities is that they regarded the ELF as their organisation and resented the treatment it received at the hands of the Eritrean government. They seem to equate the exclusion of the ELF with the exclusion of the Eritrean lowlanders, including themselves. It is as if the people from the lowlands do not count otherwise they would not be denied a say in the manner the country is ruled.

One of the key respondents put it as follows:

> "I have always been an active member of ELF. ELF means a lot to me. ELF should have been in power instead of the EPLF. The government would have not got the respect it has had it not been for the ELF. The EPLF did not liberate Eritrea on its own. The Tigray People's Liberation Front (TPLF)[9] helped the EPLF to be where it is. The current government denied the ELF the right to return because it was aware that the ELF was more historically rooted and popular than itself. EPLF is the major cause of disunity and national disharmony. Why should people return to be ruled by a government that does not care about the well-being of its own people."[10]

Another respondent said:

> "I agree with the people that say the situation in Eritrea was much better when Eritrea was under Ethiopian occupation. We knew Ethiopia was our enemy and we understood the reasons why it treated us badly. We are now oppressed by people who are members of our national community. There is not much we can do about this but continue the struggle as we did before. My father was a martyr. He died so that his children could live happily in a free and independent Eritrea. Look at what happens now. Eritrea is independent but yet not free."[11]

Many of the respondents relate the kind of expectations and dreams they had about what they would do after Eritrea became free from foreign occupation. The girls used to dream of being married to the youth freedom fighters after they had liberated Eritrea. People expected independence would create favourable conditions that would encourage people to cooperate and trust each other and consequently rebuild their country in peace and harmony. Unfortunately, all those dreams were dashed by the government's decision to create an exclusionist regime which has engendered grievance and resentment.

Most of the respondents argued that internal conflicts were integral parts of Eritrean history. They expected that independence would put an end to such conflicts. However, this could only happen in an inclusive government of national unity that represents the interests of the diverse constituents in the country. None of these expectations were realised and not surprisingly, many of the respondents rejected the idea of return to a country where not very much had changed in spite of national independence. Although the refugees argue that their dreams and expectations are shattered by the government's failure to live up to their expectation, they have not given up the hope of return to a democratic Eritrea.

None of the refugees or people I met in other settings during the field work had suffered degrading treatment at the hands of the Eritrean government. Some have been back to Eritrea for short visits and some have relatives who are jailed or detained, but since they were not subject to degrading treatment personally it is difficult to judge how the government would treat them upon a more permanent return. It is also important to point out that some of the refugees have relatives, friends and acquaintances who returned voluntarily at the beginning of 1990, some of whom have been detained.

Kibreab states that if any returnee were suspected of being an active member of the Islamic organisations, the government would not hesitate to deal with such people severely (Kibreab, 2002). In fact in the early 1990s, some renowned members of the ELF had returned to Eritrea voluntarily, but the location of some of them is unknown (Amnesty International and Human Rights Watch, reports, 2005).

Lack of Recognition

Although the need for recognition is a common concern among all human beings, it is important to those who fight for a cause. One of the reasons why Eritrean refugees want recognition is that they perceive themselves as being part of the Eritrean struggle (see Ammar, 1992). The ELF was founded by exiled Eritrean Moslems. Eritrean refugees in Sudan also contributed to the Eritrean armed struggle. The refugees generally see the factors that caused their displacement as being linked to the war of independence. Therefore, their desire to be recognised as people who matter is very strong. This was expressed by many respondents. Not only did they express a powerful desire to be recognised but they also expected the new Eritrean government to pay them a visit and express its gratitude for their perseverance and suffering, which it has failed to do. A respondent, for example, said:

> "When Eritrea became free, we expected a visit from a government delegation. Preferably, the President should

> have come with a message to say 'the conditions that forced you to abandon your country have ceased and I have come to take you home with me and to thank the Sudanese.' In response, we would have marched into a free Eritrea as free and proud people with our ELF flag. None of this happened. Instead, the leader invited us to return to Eritrea and but also urged us to forget what we had fought for. How can we return to a country that might rekindle memories of domination and colonisation? At one time the Eritrean president was an ELF fighter. He knows what the ELF stands for and what it means to its members. The exclusion of the ELF from power amounts to exclusion Eritrean Moslems, as well as the refugees in Sudan. Why does the Eritrean President visit many countries in Europe? Why not Sudan? He does this purposely. The reason why he frequently visits Europe is because his supporters are there."[12]

While I was doing the fieldwork there were frequent visits to all the refugee camps and urban areas in Sudan by the assistant commissioner of ERREC in order to inform about the conditions in the country of origin.[13] Most of the refugees do not trust the ERREC personnel. Since they do not trust the representatives of the Eritrean government, it is difficult to understand what the respondent meant by expecting the president to pay his community a visit. In view of the fact that the decision concerning return migration is a very complex process influenced by many joint and interconnected factors, it is unreasonable to imagine that the refugees in Kassala town would have marched back to Eritrea had they been visited by the Eritrean president. It is therefore important to be critical about the views expressed by some of the refugee respondents.

Some of the respondents argued that the flight of Eritrean refugees was prompted by internal and external factors. The latter is solved but the former is awaiting a solution. The party in power is part of the internal problem. After participating in the activities of a political organisation, some people develop a strong sense of attachment. Many of the respondents also reported having lost close relatives while fighting for the ELF when it was in the field prior to its ejection in 1981. For those who have powerful attachment to the ELF, its exclusion from power is not taken lightly. One respondent, for example, said:

> "The failure of the EPLF to recognise the ELF as a legitimate political organisation makes me feel that the martyrs that sacrificed their lives for the sake of Eritrea died for nothing. The ruling party wants the Eritrean people to forget about what the ELF did in the Eritrean struggle. But the truth is that the EPLF would never have come to power unless the ELF had started the struggle in the first place."[14]

Another respondent said:

> "When my father died, I was a six months old foetus in my mother's womb. I never met my father; he was martyred as a fighter of the ELF. To me the ELF is my surrogate father and I trust the organisation and those who support it. I am not sure whether I would have returned to Eritrea if the ELF shared power. I am born and raised in Sudan. However, to me the exclusion of the ELF is more personal than just political. Imagine how my father would have felt if he were alive to witness the exclusion of the organisation, he paid his life for. I do not think the government of Eritrea sees it this way. Otherwise, it would have let the organisation participate in the Eritrean political arena."[15]

Some of the refugees whose relatives sacrificed their lives for Eritrea's independence tend to take it personally when the organisation their loved ones belonged to is ignored and not recognised. They consider this as amounting to non-recognition of those who paid the highest price. Not surprisingly, this has engendered strong resentment and is allegedly reflected in the decision of the refugees to stay in Sudan.

Economic and Social Considerations

The respondents place much emphasis on issues pertaining to the political situation in Eritrea, as well as on the government's lack of respect for democratic and human rights. However, in spite of the respondents' emphasis on such issues, there is no denial of the fact that socio-economic factors also matter in the decision of the majority of the households regarding return or stay in Sudan. These are people who live from hand to mouth and in spite of the appearance the outspoken few try to portray the overwhelming majority is pre-occupied with issues of earning their daily bread. Most of them are aware of the level of devastation their country was subjected to during the thirty years war. They know reconstruction of their livelihood systems in the context of a devastated economy and infrastructure would be an extra task. The following examples may demonstrate the importance of socio-economic considerations in the refugees' decision concerning return migration.

Ali is a Moslem refugee from the Tigre ethnic group and lives in Kassala town. He left Eritrea in the beginning of 1970 with his family when his village was attacked by Ethiopia. He was a peasant from the rural areas of lowland Eritrea. He is a member of the ELF. He has never attended school although he can recite orally some verses of the Qur'an. Previously, he worked as a guard but was unemployed at the time when the interview was conducted. Both he and his wife are dependent on their two sons who have families of their own. Both sons run their own businesses. They also own their own houses. The respondent has four daughters with families of their own. According to the informant all the four daughters are relatively well off by local standards. The respondent and his wife live with the eldest son. Their grand children go to Sudanese schools and the family has access to health care.[16]

Ali's family is well integrated into the local community. It is uncommon for Eritrean refugees to own houses in Sudan because according to the Regulation of the Asylum Act, 1974, refugees are prohibited from owning property. In spite of the legal restrictions, there are many Eritreans who like Ali's sons manage to purchase houses by acquiring informal Sudanese nationality. For Ali and his family return to Eritrea undoubtedly constitutes a considerable loss of economic resources. Return would mean that they have to relinquish their businesses and sell their houses for an unknown future across the border. When well-established families like this are asked whether they intend to return to Eritrea, the answer one gets is 'absolutely yes' followed by 'when a democratic government that recognises democratic rights and respects human rights is in place.

Bli is a middle-aged single woman from the Tigrinya-speaking ethnic group. She was originally a Christian but converted into Islam in Sudan. She came to Sudan at the beginning of 1980s at the age of 19. Her parents died when she was 17. After the death of her parents, she lived with relatives on her father's side. Her relatives mistreated her and she fled to Sudan. She found a job as a domestic servant with a Sudanese family. She has no news about her family. She has no relatives in Sudan. She has some friends and the family she works for. She is also linked to members of the ELF. This means a lot to her. She was illiterate when she came to Sudan. With the permission of her employer, she was able to attend the ELF illiteracy class. She seems to value this opportunity and believes that had she stayed in Eritrea, none of these opportunities would have been available to her. She pointed out that her father was blind and the whole family was dependent on her uncle who was not that rich either.17

These two cases present different pictures. Ali is a man who has a large network of close relatives. He is a person who is part of the host society due to his socio-economic status, ethnicity and religion. Life in the host country is better than it was at home. He has a successful family network with secure livelihood thanks to his sons.18 He is well looked after by his sons. In comparison to Ali, Bli's social network is limited and she is economically vulnerable. She lacks economic and social resources. This is in spite of the fact that she has spent many years in Sudan and came as a young woman.

Ali belongs to the Tigre ethnic group which is predominantly Moslem. His family belongs to the Beni Amer ethnic group. In Kassala, the Tigre-speaking group is the second largest group and Islam is the dominant religion. Both religion and ethnicity have helped Ali to integrate into the local society. Although Ali is passionate about his political organisation the ELF, more importantly, not only does he have his whole family in Kassala town, but also all his adult children are well established. The family has too much of a stake in Sudan to accept return migration as a viable option. For Ali and other individual families in his position, no matter what reasons they give, part of the obvious reasons that motivates their stay in the country of asylum is economic. Ali has well-established sons and daughters in Sudan. Prior to his displacement Ali was a peasant. Exile has brought him and his family change of social and economic status. They have become urbanised. This means that they are accustomed to a new way of life which is completely different from their pre-flight existence. (More on the social change experienced by the refugees is found in Rogge, 1994; Kibreab, 2004).

In contrast to Ali's situation, Bli has not spent the same amount of time in Sudan. Her gender and marital status contribute to her vulnerability. Her limited scope of social networks also weakens her circumstances. In traditional societies where men and women are assigned different roles and societies give different meanings and values to such roles, female marriage is the norm. An unmarried woman is looked down upon by society as if something is wrong with her. In such societies, marriage is not seen as an end in itself but as a means to an end, the end being giving birth to children. Women are groomed from early childhood to become wives and mothers. Any woman who does not fulfil these roles is considered as inferior. Bli has to live with this social burden even though nothing of what she did or did not do was her own making.

After having spent over twenty years in exile, her community at home would expect her to have fulfilled her gender roles. Since she has not fulfilled any of the expected roles, return would mean humiliation accompanied by guilt for having failed to play the gendered roles. She was neither socially nor economically successful. Since she does not know whether her relatives are dead or alive, she has nothing to return to.

Although these two case histories are different, they show that the reasons why the refugees stay in Sudan may be very different. In both examples and probably for most refugees, their decision whether to return or not have to be weighted against a complex body of factors in their current life such as religion, political affiliation, ethnicity and economic circumstances. Some refugees' return is discouraged by economic success whilst for others, economic failure accounts for a non-return option.

Sudan is a much larger country than Eritrea and being endowed with the perennial rivers of the two Niles and the Atbara, there are greater opportunities for income-generating activities, employment and trade than in Eritrea. Thus, even for vulnerable individuals such as Bli life in exile has not been that bad.

State Ownership of Land in Eritrea

In Eritrean society, the importance of land is not solely measured in terms of economic return. All Eritreans are rooted in particular places. That is the reason why land is a source of social identity. A person without land is regarded as rootless. This was what elderly refugees, male and female, told me many times in my discussions and interviews with them. They did not put it in these exact words but they explained the close relationship between the person and the land from which s/he comes.

In 1975 the Ethiopian military junta, the Derg, nationalised all land throughout Ethiopia, including Eritrea. However, since most of the Eritrean rural areas were controlled by the liberation movements, the Derg's policy on land had no impact on traditional Eritrean systems. De facto all land continued to be owned by the various ethno-linguistic groups.

In 1994, the post-independence Eritrean government enacted Proclamation 58/1994 and vested upon itself the ownership of all land throughout the country. Not only did the government abrogate the diverse traditional land tenure systems but also the customary laws that regulated access to and use of land were also repealed. Land users have only rights of usufruct on land owned by the state (Kibreab, 2005). The data elicited from key refugee respondents show that many are against the policy of land ownership in Eritrea. Upon return, they want to regain the land they left behind when they fled. Some of the respondents maintain that without such a change in land policy, those who returned would have been able to reclaim their old possessions, but now they have to seek permission from the government to do so. They were not sure whether their old possessions were taken over by others after independence. Some of the informants reported that the issue of land is one of the factors that have discouraged them from returning.

Since all land in Eritrea belongs to the state, farmers only have rights of usufruct or use. In Eritrean society land is a source of identity, pride and status. It is also the single most important source of livelihood for the large majority of the population. The decision of the government to invest upon itself the ownership of all land by disregarding its meaning and significance to the Eritrean people has engendered resentment, particularly among those who originate from the Western lowlands. Although the whole country is affected by the land policy of the government, the refugees from the Western lowlands believe rightly or wrongly that the land policy was adopted to benefit the

Christian highlanders at the expense of Moslem lowlanders, particularly the inhabitants of the Baraka sub-region.

Unlike in the Eritrean highlands where there is a problem of overcrowding, there is abundant land in the Western lowlands. The respondents from Baraka argue that the reason why the government took over ownership of the land was to resettle Christian highlanders in Baraka where there is plenty of cultivable land.[19] One of the key respondents, for example, said that when he returned to Eritrea he found out that the highlands are inhabited by the same population as in the past whilst the demographic composition of the western lowlands has changed considerably due to the influx of highlanders. In his view, the main reason the government-nationalised land was to facilitate the settlement of highlanders in Baraka.

The respondent pointed out that the government has expropriated land belonging to the people of Baraka in order to transfer it to Tigrinya speakers. He also pointed out that before Eritrea's independence; the EPLF had its base area in Sahil. The people of Sahil helped the EPLF, but after it took over power the latter has neglected the needs of the population of Sahil and has been solely concentrating on the development of the Tigrinya speakers. This respondent pointed out that he was not against the people from the Kebesa (highlanders). He further said that he did not mind if people from the Kebesa or Keren were settled in Baraka, but he was against the government's expropriation of land. He pointed out that in the Past, it was common for people from these areas to come to the Baraka area as migrants, not as landowners as is the case at present.

> "We the owners of the land are in exile whilst the government distributes our land to the highlanders. The government says that it wants us to return, but if it really means it, it should have allowed the ELF to share power. Only then would our return become possible."

It is interesting to see how the respondent after arguing

that land was a major issue, says that the owners of the land (with whom he identifies) would have returned if the ELF shared power even in the context of state ownership of land. The respondent also accused the government of stealing the land belonging to his community and for not caring about the refugees in Sudan. The reason why it did not care, he thought, was because the remaining refugees are mainly Moslems, which the government does not consider as its own. He did not think that any Moslem would want to return to such situation. He said:

> "In Sudan we are living a respectable life. Although it is not our country, we have our dignity. What is "*watan*" (a country) without dignity? People need to feel that they are part of society, being part of society is participation in the society on equal terms. Being marginalized or being dominated is not a reflection of equal treatment."[20]

Because the respondent doubts the intention of the government, he construes all its policies as being anti-Islamic and anti-lowlanders. This particular respondent is a Moslem himself and he is a member of the ELF. His formal explanation as to why he has to stay in Sudan is the Eritrean government's land policy and the exclusion of the ELF. His statements are also coloured strongly against the Eritrean government. He refers to himself as 'we' as if he has been delegated by the Eritrean refugees to speak on their behalf.

Respondents like the one just quoted are very angry with the government. Therefore whatever they say about the latter cannot be taken at its face value. It is important to critically evaluate the validity of their comments. The respondent gives the impression that had it not been for the exclusion of the ELF, he would have returned to Eritrea. Perhaps return to Eritrea is not an attractive option to him and to his like. Because the ELF is excluded from power, the respondent's allegation can be to some extent politicised. This is one of the methodological problems I faced in carrying out this research.

This respondent and others like him give the impression that the government favours Christian highlanders at the expense of Moslem lowlanders. There are a number of problems that arise from such blanket generalisation. The term Moslem lowlanders give the impression that the inhabitants of the Eritrean lowlands are exclusively Moslem. This is incorrect. There are many Christians who regard the Eritrean lowlands as their homeland.

The geography and history of Eritrea did not begin in 1890 with the coming of the Italians. The land called Eritrea was settled, and the inhabitants had traditional and customary ownership to their landed property. In Eritrea there were three different kinds of land ownership: the *Resti* (Tselmi) family ownership, an inherited land ownership from generation to generation, the *Deisa* village ownership which circulates every 5–7 years between the people of the village (only men), and *Community* ownership (tribe or clan), mainly in the eastern and western lowlands with a few exceptions in the Gash area. In the community owned land each tribe or group has three main places which are private; these are the winter accommodation villages called Dammar, here they build stable cottages for their families, the second is the winter place Shillbo, their houses here are movable and temporary. The third place is the water place, a well or running water during the winter and always a place with big trees under which their cattle can rest after drinking water.

This was before the coming of the Italian colonizers who confiscated all the land in the lowlands and introduced *Demoniale* government ownership. Demoniale was a means of seizing the fertile land near the rivers Baraka and Gash and giving it to the Italian settlers. They introduced mechanized farming by irrigation, the Ali-Geder cotton project is an example. However, the pastoralists and nomads did not change their traditional way of life, for example Eritrean herdsmen from Baraka and Gash crossed the borders via Setit to Ethiopia and camel men crossed the borders to Sudan. The Sudanese did the same reaching the Eritrean highland. Nevertheless, throughout the history of Eritrea land remained in the hands of its traditional owners. The only time that Eritrean people lost control over their land is after Eritrea's independence in 1994. As the rest of African countries, Eritrean society is agrarian. People's survival is based on farming. Therefore, land ownership is a vital, which can cause internal conflict among Eritreans if not tackled in the right way. This was something of which my respondents, in particular the elderly ones, were acutely aware.

The Italian government enacted a law in 1909 and 1927 that transferred all land in the Eastern and Western lowlands to state ownership. These laws have never been repealed since then. Hence, at the formal level, the Eritrean government has not radically changed the land policy in the Western lowlands. This does not, however, mean that the people should not expect an independent Eritrean state to return land to its rightful owners. It is unfortunate that instead of reversing the injustice imposed by external forces, the government continued along the same line.

Such respondents also give the impression that Eritrean lowlanders have common interests and common goals as if they are not differentiated by class, ideology, ethnicity, politics and religion. During the last century, the Eritrean Western lowlands have undergone considerable changes demographically, physically and socially and these changes cannot be wished out of existence.

If the refugees have undergone change in Sudan, it is unreasonable to expect stasis in their areas of origin. The other question, which arises in connection to some of the respondents' complaints, is the question of whether every national has the right to live anywhere within Eritrea. The expectation of some of the refugees to exclude nationals from the other parts of the country is unreasonable. As is generally the case, every national has the right to settle legally anywhere within the territories of Eritrea. Inasmuch as people from the highlands have no right to say that people from the lowlands cannot settle in the highlands, the inhabitants of the latter similarly have no right to say 'no' to the decision of people from the highlands to settle in the lowlands if they choose to do so.

Kibreab's study on the relationship between returnees and the stayees in Eritrea shows that the two groups live in harmony with each other. This was true regardless of the returnees' religion, ethnicity and place of origin. There is yet no empirical evidence to show that the local population in the receiving areas is hostile to 'outsiders' who settle in their areas (Kibreab, 2002). Although respondents such as the one quoted in the preceding paragraph give the impression that the effect of the government's land policy on all Eritrean lowlanders is uniform, it is important to point out that those who previously did not own land are likely to view the new policy positively because they would be able to have access to government owned land.

However, this would only be true if the government's policy benefited all Eritreans. Several respondents21 in Kassala town believe that the people who are benefiting from the government's land policy are individuals who made money in the Gulf States and who support the government. Whether this belief has any foundation in actual facts is impossible to tell. However, it does show that the Eritrean government's land policy is highly controversial, which provides nourishment for rumours about who stands to gain and who stands to lose.

Religion, Ethnicity and Language

Eritrean society is deeply religious and unless the question of religion is handled with care, it can be a cause of polarisation. During the war of independence, the EPLF was aware of the sensitivity of this issue. That was why it declared Wednesday rather than Saturday and Sunday as a weekend. Friday is as important to Moslems as Sunday is to Christians. Since Wednesday was a neutral day, both Moslems and Christians accepted it as a reasonable compromise. During the federation, Arabic and Tigrinya languages were national languages and Friday and Sunday was regarded as a weekend in which all government employees rested. These polices were based on the Eritrean socio-religious culture.

After the abolition of the federation, Eritrea became a province of Ethiopia, Amharic became a national language, and national holy days were Saturday and Sunday. The Eritrean people did not protest openly against the new work free days for employees in the formal sector. The strange thing is that even after Eritrea's independence Friday became a regular working day as it was before. This was deeply resented by some respondents[22]. Justifiably, most Eritrean Moslems expected the post-independence Eritrean government to reinstate the status of Friday. However, instead of treating Friday and Sunday equally, the Eritrean government followed the discriminatory precedent set by the Ethiopian government. Friday is a normal working day. The refugees interviewed in this study construe this as representing a discrimination against Moslems. Several of the respondents pointed out this as evidence for the government's unequal treatment of Moslems.

In an attempt to understand the rationale of the Eritrean government, I interviewed the staff of the Eritrean Embassy in Khartoum in 2002. In the interview, I was told that the government was studying the issue and the policy was expected to change. That was in 2002 and the policy has still not changed.

Language is another contentious issue. Although the Eritrean constitution states that Arabic and Tigrinya are both the working languages in the country, the refugee respondents argue that Tigrinya is the dominant language and Arabic is neglected. Some respondents linked their reluctance to return to the government's discrimination of Arabic. Although it is difficult to say with certainty the extent to which the refugees' decision is negatively affected by the government's language policy, there is no doubt that this is of a concern to the refugees.

In Kassala, most of the children belonging to Moslem refugee families are taught in Arabic. For these families whether their children carry on with their studies upon return is to a large extent dependent on the use of Arabic as a medium of instruction in Eritrean schools. In Eritrea, the government has opted for mother tongue language policy, i.e. all ethnic groups are taught in their first language at a primary level. There are also primary schools, especially in the areas inhabited by Moslems that use Arabic as a medium of instruction. This is invariably the case in the areas of return. In post-primary school education, English is the medium of instruction and Arabic and Tigrinya are taught as languages. In Sudan, Arabic remains a medium of instruction even at a university level except in the faculty of medicine. Hence, the refugee families are concerned about the difficulties their children may face upon return. Generally, for the Moslem community the reinstitution of Arabic as one of the two national languages is also important. Studies in Eritrea (Naty, 2002) show that Moslem communities in Eritrea oppose the policy that schoolchildren peruse their respective mother tongues to attend primary school (grade one to five) because they regard the pursuit of such a policy as being conspiratorial. In the view of those who are against the mother tongue policy of the Eritrean government, some of the languages do not have scripts and therefore are unable to compete with Tigrinya and therefore the latter will continue to dominate the scene.

The fact that the government's working language has in effect become Tigrinya is also putting Eritreans who study in the Middle East at a disadvantage when they return home. The labour market, especially the public sector becomes almost inaccessible. The following example clearly demonstrates this problem.

"After independence, I returned from Sudan to Eritrea

to look for job. It took me two years to get one. After I applied for a job, it took the authorities seven months to reach a decision. I was informed by people in my profession (engineering) that I should be fluent in English and Tigrinya otherwise it would be impossible for me to find a job. That is the reason why I have come back to Sudan. Now I work with an NGO. The highlanders, Christians and EPLF control Eritrea. There is no room for the rest of the people unless they try to be like them. People in Sudan talk only about politics, but the government controls everything. We Moslems are excluded. In fact, the Moslems who are in government live the same life as that of the highlanders. They speak in Tigrinya and consume alcohol. It is not easy for Moslems to apply for business licence either. For the highlanders, it is easy to obtain a business licence. However, if you to open a bar a licence will be issued to you promptly. During the Ethiopian regime, women working in the bars are prostitutes. Many Moslem women have been given jobs in this branch. When Eritrea was under the colonial regime, things used to work differently. It was unthinkable for Moslem women to work in bars. However, nowadays it has become common for members of the Moslem community to engage in such activities."[23]

There are a number of issues raised by this respondent that deserve critical analysis. The problems he faced upon return are commonly faced by Eritrean returnees from Sudan and the Middle East where the medium of instruction is Arabic. Moslem Eritreans mostly face this problem because they were trained in Arabic. Thus, it is legitimate to blame the problems the respondent faced upon return on the government's wrong language policy.

However, the respondent did not only complain about issues related to work. He said that Moslems in Eritrea are adapting a highland lifestyle, e.g. speaking in Tigrinya and drinking alcohol. It is clear why the respondent found this irritating. It seems that the respondent disapproves of alcohol consumption for ideological reasons. However, he forgot that all Eritreans have the right to choose their life styles. If the respondent resents government imposition, it is unreasonable for him and for others like him to try to impose their values on other Eritreans.

His position is shared by many exiles who scream for democratic rights. This poses some questions as to what their perception of democracy is. Generally, the idea of democracy requires the right and freedom of individuals to choose the life style they lead. Would refugees such as the one quoted above be prepared to subscribe to these freedoms of the individual in a democratic society? Would they, anymore than the present government recognise the choices individual must be allowed to make in their personal lives?

It is common for the refugees to speak in polarised terms as 'us' and 'them.' Most of the people interviewed are Moslem lowlanders the large majority of whom are from the Tigre. The Tigre has its own ethnic group in Kassala. The majority of the inhabitants of Kassala are from the Beja ethnic group and the Tigre speaking Beni Amer are the second largest group. It is possible that some of them feel at home in Kassala and that most of them have acquired informal Sudanese citizenship. Their failure to return to Eritrea in spite of their reported reasons could quite possibly be due to the fact that they are well integrated into Sudanese society and see no compelling reason to change this.

Gender Roles and Exile

In Eritrean society, nearly all aspects of life are gendered. Gender inequality permeates every social, economic, cultural and political aspect of the communities regardless of ethnicity, religion or region (Kibreab, 2003). The roles played by men and women and the meanings and values the communities attribute to such roles are socially constructed or learnt (ibid.). However, these roles are presented as if they emanate from men's and women's biological differences. Men are perceived to be the breadwinners of their families, as well as independent, emotionally mature, sexually dominant and endowed with certain natural qualities that make them superior to women and children (Kibreab, 2005).

On the other hand, not only are women's innate qualities perceived to be deficient emotionally, they are also said to be dependent on men, not only for their livelihoods but also for their physical security. Although Eritrean society is diverse and consequently the formal and informal rules that govern the relationship between the two sexes are different, in nearly all groups, there are gender-based divisions of labour governed by gender-based ideology that defines what is regarded as appropriate to do or not to do for a person of a given age, gender and status. These rules are adhered to even by those who are supposed to know better as if they were God-given.

Yet in most of the communities, women participate in production as well as in domestic work. Cooking, cleaning, laundering, food processing, child rearing and minding, nurturing of men and children, nursing of sick and elderly family members are exclusively the responsibility of women. In addition to all these responsibilities, women in rural areas also take an active part in productive activities such as weeding, harvesting, transporting products from farms, etc. (Kibreab, 1995). For most Eritrean women, the working day tends to be very long. Worse still, the role they play in the maintenance and reproduction of their families and communities is not recognised. Every female is socialised from early childhood to become a good mother and a wife. Although the survival of a family may depend on an income earned by a wife, the husband in spite of his failure to contribute to the subsistence needs of the family concerned may still be considered as the breadwinner of the family (Bascom, 2005). The notion of breadwinning is a socially a constructed concept which in most cases may not correspond with the reality on the ground (Kibreab, 2003).

In Eritrea, newly married women live with their in-laws until the husband becomes economically independent. Women look after the family home, children, elderly and sick members of families. In the marital period, when the newly wed women are required to stay with their in-laws, they become subordinate to their whims. Eritrean society is highly patriarchal (Mcspadden, 1999; Favali & Pateman, 2003). Patriarchy is a system in which men dominate, oppress and exploit women (Walby, 1990, p. 20; Duncan, 1994).

Gender relations like any other phenomenon is subject to change and transformation. These processes of change and transformation are often accelerated in crisis situations. During the war of independence, the traditionally prescribed gender roles were to some extent undermined by the new roles many women played during the thirty years war of independence. Women played a key role in the two liberation movements during the difficult war years. This was especially true about their participation in the ELF and EPLF. It is not clear the extent to which Eritrean society has changed its view of women as a result of the role they played during the war of independence. According to discussions24 with Eritrean women who were freedom fighters, after independence, society expected women, including those who played a vital role during the war of independence to return to the kitchen

This is notwithstanding the fact that the Eritrean government has formally argued that it abolished all the laws that discriminate against women (Tekle, 1998). According Kibreab this does not however suggest that in reality they are treated equally. As he, observes:

> "Though the legal reforms introduced by the government of Eritrea and the measures introduced during the war of national independence are important, it is vital to realise the existence of the wide gap between the law and the reality on the ground. This is important because no matter how progressive laws on gender relations might be in the law books, the reality on the

> ground is what counts in the peoples' daily lives and livelihoods. Legal reforms on gender relations though important per se are not decisive unless they are enforced rigorously and complied with effectively." (Kibreab, 2005, p. 12)

An important factor apart of the national law is of course the customary and Islamic laws that govern personal lives. To what extent are these laws subordinated to the equality declared in the national constitution?

Forced migration is another factor which has considerable transformative impact on gender relations (Indra, 1991). Forced migration has to some extent led to change and transformation of gender relations among the refugees. In many cases, displacement destabilises the socio-cultural and economic structure on which the inequality between men and women is based. One of the critical factors that contribute to women's subordination to men is the actual or perceived breadwinning role played men. Men's breadwinning role becomes undermined in the immediate post-flight condition when they fail to protect their families against attack, hunger and thirst. Even after arrival to safety, there are studies to show that it is women who bounce back to meet the basic needs of their families by seizing every opportunity to earn an income and to rescue their families from starvation and humiliation. Gaim Kibreab has documented how female Eritrean refugees in Khartoum disregarding their pre-flight socio-economic status, accepted the most menial jobs such as domestic service in Sudanese households, while their husbands wasted their time mourning about their lost status and prestige (Kibreab, 1995).

As a result, previously naturalised roles played by men and women became increasingly blurred as more women played the breadwinning role in the communities. In exile, usually the old gender-based divisions of labour become weakened and this enables some women to renegotiate the rules that govern intra-household relations. This new situation tends to lead to new social arrangements. This was, however, only true in urban areas where no outside assistance was available and employment opportunities for women in Sudanese families were available (ibid.).

However, it is important to realise that gender relations are dynamic and this is reflected in the changes that occur in connection with involuntary displacement, it is equally important to realise that the relations of inequality do not disappear easily. In the case of Eritrean refugees in Kassala, there have been some changes resulting from the fact that women are earning their own incomes and contributing to their families' subsistence needs. The traditional stereotypes about what women ought to do or not do appear to be increasingly challenged.

In spite of the changes, even in the families where women are the sole breadwinners of their families, an adult male member is invariably considered as the head of the household (Kibreab, 2003). This has considerable consequences because it is often the household head that makes the important decisions that affect the well-being of the family members.

It is interesting to note that among the refugees covered in this study the impacts of migration on gender relations are reduced by the local patriarchal norms that prescribe that a woman regardless of her profession, education or status needs a male guardian. Before marriage, the role of guardianship is played by a father or brother and after marriage by a husband and in his absence by an adult son, or the spouse's father, brother or uncle.

In spite of these, there are still some changes that have had an impact on the views of gender roles among the refugees in Kassala. Data drawn from female respondents show that some kinds of changes have occurred because of involuntary displacement. Some of the factors that contributed to such changes include the absence of men in many families.

Many men immigrated to the oil-rich Gulf States leaving some members of their families in Kassala town. In many cases those family members who remained behind and who live off remittances sent by their relatives are also headed by women. The impact of this on pre-existing gender relations appear to be considerable. The following testimony by a female respondent may demonstrate the significance of income-earning capacity of women on pre-existing gender roles. The respondent recounted:

> "I am playing the role of a mother and father. I do what a father does outside the home and what mothers do inside the home. According to the code of my clan, women who played such a role, i.e. those who transgressed the socially constructed boundaries were stigmatised."[25]

The change in gender relations that have occurred among the refugees is found to be not only reflected in the changing roles played by women but also in the attitudinal change this is engendering among the refugee communities. In her testimony, the respondent recounted that in her clan, a woman who played a breadwinning role, i.e. working outside the house to meet her family's needs by transgressing the social boundaries was stigmatised and in extreme cases ostracised. In Kassala, women who work in order to meet the needs of their families are no longer stigmatised. Of course, it is important not to overstate the extent of attitudinal change that has taken place among men. In Sudan as well as Eritrea it is a common norm to regard women as being subordinate to men, there are still very powerful groups in the refugee communities who resent women's independence and would want to confine them to the compound. What is interesting and this can be interpreted as a reflection of some degree of empowerment is that women are disregarding the views of men by rejecting their expectations in order to meet the needs of their families. There is very little that the conservative forces can do to arrest the process of social change that is taking place in the area of gender relations. The respondent further pointed out:

> "In Eritrea, i.e. prior to our displacement, I was dependent on my husband's handout. I depended on what my husband gave me to buy clothes, perfumes or to buy gifts to relatives and friends. I took care of the home and he took care of whatever happened outside our home. I was only responsible for domestic work. I just cooked, listened to the radio and watched TV."

The confinement of women or their relegation to the domestic sphere does not only deprive women of income and the social benefits that are derived from the capacity of earning an income but this has considerable impact on women's well being and experiences. The important question to ask in connection to the change in the breadwinning role played by women is whether the particular women who are undergoing these changes celebrate or regret these changes. As outsiders, we tend to welcome such changes, but for the women who have to bear the burden of the tasks and activities, the changes may represent an unbearable burden in the short-term. In an attempt to understand the perception of the affected women, I asked them how they felt about their new experiences and responsibilities.

For one of the respondents who referred to herself as the 'the man of the house,' the changes were not necessarily worth celebrating. They had increased her burden. The fact that she referred to her new role as 'the man of the house' shows the extent to which she had internalised and naturalised the learned roles played by men and women in her society. This may show that the capacity to earn an income if not accompanied by change in social and political awareness of those who undergo change may not necessarily lead to change in the relations of inequality between men and women. This respondent thought that her life was 'not normal' as she was forced to play the role of a man. The reason why she thought her life was not normal was because unlike the other 'lucky women' she was doing a man's job. The fact that she referred to the women who were subordinate to their husband's and lived off handouts of their husbands as 'lucky' clearly indicated she regretted rather than celebrated the changes in gender relations. It is important, however, not to generalise this particular respondent's testimony to the rest of the women who have undergone change and are playing a breadwinning role in their families. In the words of the respondent:

> "My three daughters are twelve, eight and ten years old. My son is fourteen. They all go to school. After school hours, the two older daughters work as babysitters and my son is an apprentice in a goldsmith workshop. In my culture, women prefer marriage to being unmarried. It is better to be married to any man than being unmarried. Having a man around the house is very important. After my husband's death, I decided to remain single. Although I never thought that my brother would mistreat me as he did, I managed to remain single for the sake of the children. Raising children as a single mother is daunting, but I have got used to it. I hope my son will take over as the father of the family once he grows up. As you see, I raise my children by selling tea. I also wash clothes on Fridays. I was 26 years old when I lost my husband. Very few women in my position would refuse to be remarried because marriage is a source of respect for women. The family of my husband returned to Eritrea after Eritrea became free. I refused to do so because I was unsure whether I would be able to continue my work and raise my family. I came to Sudan at a very young age. From the way I hear people talk about women who work outside the house, I don't think it would be possible for me to work and earn an income as I am currently doing in Kassala."[26]

There are a number of interesting observations one can make in regard to the respondent's testimony.

Firstly, although she is struggling to meet the needs of her family by transgressing the roles played by women in her family, she would rather not play these roles if she had a choice. In fact, she was looking forward to the day when she would relinquish her headship and breadwinning role.

That is what she meant when she said she hoped that one day her son would be a man and a father of the family, which then would include herself. This is in spite of the fact that she has been playing head of the family for several years and her son was at least four years away from manhood. This particular woman and like-minded others must be suffering from some sense of guilt resulting from their lack of 'respect' to the socially prescribed gender roles.

Secondly, her testimony also shows that among the poor refugee households, the social norms that are prescribed by religion and culture are ignored under the pressure of the need to eke out a meagre existence that suggests that there are economic thresholds beyond which cultural norms are not complied with.

Thirdly, her two daughters who are engaged in part-time income-generating activities may not over time share their mother's view about gender-based divisions of labour. Because they belong to a different generation and have begun to be engaged in income-generating activities at an early age, by the time they grow up, they may reject the notion of women's subordination to and dependence on men. Education is also likely to contribute to their political and social awareness. However, there is a qualification that ought to be borne in mind. In view of the fact that their mother is advocate of women's marriage and believes that any marriage is better than no marriage, she may force her daughters into male subordination and dependence. It remains to be seen whether the girls will resist or succumb to their mother's and their relatives' social pressure.

Fourthly, women who reject the social norms that prevail in their communities risk exclusion. This particular respondent's cries for the ill treatment she received at the hand of her brother. It seems that she expected her brother to come to rescue her when she faced difficulties after the death of her husband. Soon after her husband's death, her brother-in-law wanted to inherit her by making her his wife. Since this was a common practice in her clan, her family expected her to meet the terms happily. She rejected her brother-in-law's marriage offer. Her family urged her to reconsider but she refused stubbornly. It was in response to her 'unacceptable behaviour' which defied long-standing tradition that led to her excommunication from her family. After her rejection of the offer, her brother and other relatives did not want to hear from her. She was left on her own in the cold. That was why she had to go out and sell tea in the market to meet the subsistence needs of her family. It is not clear why she turned down the marriage offer. Probably she is against the practice of wife inheritance. It could also be because she did not love her brother-in-law. It could also be due to respect for her deceased husband. Whatever the reason her decision to trim against the wind shows the dignity and independence for which she paid a heavy price, namely, familial support. It is surprising to see her expressing great respect to the institution of marriage in spite of her decision to reject it. It seems that she values marriage and thinks married women are more respectable than unmarried women, provided the decision is based on mutual consent. She is definitely ahead of her time reflected in her decision to stand by her belief in spite of the social and economic costs involved. It is not clear the extent to which she would have been able to do this had she not been exposed to the new experience associated with involuntary displacement.

The importance of this respondent's testimony, among other things, lies in the fact that she took the decision to remain in Sudan instead of following her family to return to Eritrea for fear that she might not be able to carry on with her occupation. The main factor that discouraged her was the stigma attached in her clan to women's income-generating activities that takes place outside the family home.

The respondent seems to harbour a sense of shame in having violated her clan's expectation by rejecting the common practice of wife inheritance in favour of earning her own and her family's subsistence needs. Such pressure may be stronger in the country of origin than in the country of asylum. This indicates the complexity of the factors that encourage or discourage return to the country of origin of the refugees after prolonged duration of exile. Moreover, this testimony indicates the influence of migration on gender roles.

Open-ended Military Service

Another important factor that has been forcing people to flee in the post-independence period and which has been discouraging those in Sudan from returning is the open-ended military service in the country (Human Rights Watch, 2005). In 1994 the Eritrean government introduced a military service which required all citizens, including women between 18 and 40 years, to participate in national service, which entailed six months military training at Sawa military camp and one year participation in national reconstruction. Since the border war broke out in 1998, the national service has become an open-ended obligation. Most of the young refugees I interviewed in Kassala town were not in principle against the national service. They were only opposed to the indefinite extension of the obligation and against women's participation.

One of the former participants of the national service who fled to Sudan, for example, reported as follows:

> "I joined national service in the beginning of 1996 and stayed there until 1999. The duration is 18 months but many participants were in the service for many years more. After we completed two years, we were not allowed to leave. In the military training camp, the participants, especially women suffered from different kinds of harassment. The government did not take any action to stop the abuses and harassments and therefore no one was willing to stay there. Many students who were willing to continue their studies have been kept in Sawa against their will. The open-ended national service is the main cause of famine in the country. As long as the current regime remains in power,

> the country will continue to suffer under its control. I love my country and I want to return provided there is a change of regime. In the absence of such change, I have no plan to return or to stay in Sudan. My intention is to emigrate abroad. Unfortunately, I may not able to realise my dream because I have no relatives who can facilitate my emigration to the oil-rich Gulf States or Western Europe. I hope the UNHCR includes me in its resettlement programme to Canada or Australia."[27]

As seen earlier, women played a key role during the war of independence. However, the difference between the national service and the war of independence is that in the former women joined the struggle out of their free will without being forced to join the ELF or the EPLF. In the national service, women are forced to join against their will. Most parents rejected the government's policy regarding women's participation in national service. Resentment is common amongst parents who argue that this violates the code of custom. The government carries on regardless. The attitude of parents regarding national service is also influenced by allegations of sexual abuse at the hands of senior military officers. Whether this is true or false does not seem to matter. The common perception is that practices of abuse exist and this intensifies parental disapproval.

Conclusion

One of the factors that this study aims to examine is the political question regarding return migration. As indicated in this chapter and from the history of Eritrean politics it is clear that recognition and political inclusion are important factors when/if refugees consider return migration. Consequently, the exclusion of the political factor in the policies of return migration and understanding the decision the refugees make, would be a mistake. The regime excluded the ELF from power. However, the ELF is not only a political organization, because for many of my refugee informants it serves as a substitute family. Moreover, from the findings of this study the ELF is a faction that a majority of Eritrean Moslems as lowlanders are associated with or feel affinity with. Then what kind of indication does this give us? What may appear to be political in one instance may another time be social and vice versa. "Home" to these respondents is where their political faction is present. Their loyalty to the faction is explained by the fact that they feel that their faction protects their interest. The changed gender roles resulting from migration are in this case not acknowledged as something positive, which departs from the results of other migration studies.

NOTES:

1. One of the participants in interviews and discussions conducted with male merchants in Kassala market, 2002.

2. Discussion with second generation women in Kassala market and Muraba'at, October 2002.

3. Family man who left Eritrea in 1979, worked for the local transport company in Eritrea, an ex-member of the ELF, lives in Muraba'at, works in Kassala market, finished junior high school. He belongs to the Tigre clan and is a Moslem. The interview was held in Arabic and translated into English by me, Kassala, 2002.

4. Discussions and interviews with refugee groups in Muraba'at, Kassala, 2002.

5. Female, who left Eritrea in 1975. She is single, a Christian and beliongs to the Tigre clan. She I a university graduate, works with an NGO, lives in Muraba'at and is a member of the ELF. The interview took place at the respondent's work place and was held in Tigre. The transcript was translated into English by me, Kassala, 2002.

6. A family man who left Eritrea in 1981. He works with a NGO, belongs to the Blin clan, is a Christian, junior high school graduate, and an ex ELF member. He resides in Abukamsa. The interview took place at his place of work. It

was held in Blin and translated into English by me, Kassala, 2002.

7. Personal communication with a group of informants in Kassala town. The meeting was held in the home of one of the refugee families in Muraba'at. The discussion was in held in Tigre. Kassala, 2002.

8. Family man who left Eritrea in 1978. He is a senior high school graduate, a Moslem, ELF supporter and belongs to the Blin clan. The interview took place in his family house. It was conducted in Blin and translated into English by me, Kassala, 2002.

9. Tigray People's Liberation Front.

10. Female who left Eritrea in 1975. The interviewee is single and illiterate. She belongs to the Tigrinya clan and is a Moslem. She resides and works in Muraba'at. The interview took place at her working place and was conducted in Tigrinya. The transcript was translated into English by me, Kassala, 2002.

11. Male, single, born and raised in Kassala. He is a university student, belongs to the Tigre clan and is an ELF supporter. The interview took place in Abukamsa and was conducted in Arabic. The text is translated into English by me, Kassala, 2002.

12. Interviews and discussions with men in the Kassala market. The conversations were held in Blin and Tigre, I translated the text into English, Kassala, 2002.

13. Some of the refugee committee representatives said: "ERREC staff are undercover agents of PDFJ" (the ruling party in Eritrea). In an attempt to understand the rationality of this I interviewed former ERREC staff, who used to work in Sudan. In Eritrea this person worked as a security officer but was sent by the government of Eritrea to work with the ERREC staff in Sudan. It needs to be mentioned that a well-known officer of Amnesty international asked me for detailed information about the involvement of ERREC in the Sudanese refugee repatriation programme.

14. A Moslem housewife and a mother of seven, personal com-

munication. This respondent left Eritrea in 1970. She belongs to the Blin ethnic group. She lives on remittances from her husband who lives in Saudi Arabia. The interview took place in her friend's house in Abukamsa and was conducted in Blin. I translated the text into English, Kassala, 2002.

15. A male, born and raised in Kassala. This respondent is single and he lives with his single mother. He is a student of medicine and belongs to the Tigre clan. The family receives support from the interviewee's uncle who lives in Khartoum. The interview was took palace in a teashop and was conducted in Arabic and translated into English by me, Kassala, 2002.

16. Male respondent who left Eritrea in 1982. He is single and has finished senior high school. He belongs to the Tigre clan and works in the transport sector. The interview took place in the Kassala market under the shade of a tree. It was conducted in Tigre and translated into English by me, Kassala, 2002.

17. Personal communication. The interview was conducted in Tigre and was translated into English by me, Kassala, 2002.

18. Access to employment, education for his children, clean water and electricity, access to health care and to residence etc. Kassala, 2002.

19. Since some of the respondents, are the first generation and they are dependent on their children for economic supply, their argument is social and political. The respondents negative reaction is pointed at the regime as responsible for the land policy.

20. Family man who left Eritrea in 1979. He is illiterate. He is now retired and financially supported by his sons. He belongs to the Tigre clan, is a Moslem and an active member of the ELF since 1965. The interview took place in his place in Murarba'at, and was conducted in Tigre. The text was translated into English by me, Kassala, 2002.

21. One of the most outspoken was a housewife who left Eritrea in 1970. She is a Moslem, belongs to the Blin clan and is an ex-ELF supporter. She is illiterate and lives on remittances.

The interview took place in her house in Abukamsa and was conducted in Arabic. It was translated into English by me, Kassala, 2002.

22. Among them is Osman, who is a family man. He completed senior high school in Eritrea, which he left in 1975. He is a Moslem, an active member of the ELF and chief of Tigre clan. He works with an NGO. The interview took place in Muraba'at in his car, and was held in Tigre. The interview script was translated into English by me, Kassala, 2002.

23. Male, left Eritrea in 1978. He is single and a senior high school graduate. He belongs to the Tigre clan and sympathises with the ELF. The interview took place in the Kassala market and was conducted in Tigre. It was translated into English by me, Kassala, 2002.

24. In Kassala, June, 2003

25. A single mother and head of household who left Eritrea in 1977. Shehas finished junior high school. She is a Moslem, belongs to the Blin clan and resides in Muraba'at. The interview took palace in Abukamsa at her friend's house. It was conducted in Blin and translated into English by me, Kassala, 2002.

26. Female, head of household who left Eritrea in 1958. She is illiterate, belongs to the Tigre clan and is a Moslem. She resides in Muraba'at. The interview took place in her working place in Muraba'at and was held in Tigre. It was translated into English by me, Kassala, 2002.

27. Tigirnya speaker, University student, Christian, left Eritrea at the beginning of 2000, the interview was conducted in his place and was held in Tigrniya and translated into English by me, Kassala 2002.

6

THIRD-COUNTRY RELATED FACTORS
The Dream of Resettlement

Most Eritrean refugees that I met in Sudan even those without experience of urban life dream of joining the Eritrean Diaspora instead of dreaming about returning home. There are a number of reasons for this. Firstly, the government of Sudan does not prescribe to the policy of refugee naturalisation. And according to the UNHCR policy the refugees are expected to return home when the conditions that forced them to flee come to an end (Kuhlman, 1994; Kibreab, 1996). This is regardless of the length of time the refugees have been in the country. This policy induces the refugees to seek alternative solutions and resettlement to the countries of north and Australia is the most attractive option. Secondly, many of the refugees in Kassala have members of their families in either Australia or countries of north. The dream of most of those who are in Kassala as seen before is to join their relatives abroad. Thirdly, there are also refugees whose safety and security cannot be guaranteed in Sudan. The UNHCR seeks countries of asylum for such refugees where they can live in dignity and safety. For such refugees resettlement in third countries of asylum represents a durable solution (Akuei, 2005). However, the opportunities for resettlement are very restricted and only a few make it out of thousands.

The aim of this chapter is to discuss the factors that are related to resettlement to third countries and that have a bearing on the refugees' decision concerning return migration. The chapter is based on data derived from interviews with key informants and group discussions that took place during coffee and tea visits to Eritrean families in Kassala. It addresses the question why many of the refugees in Kassala town want to migrate further rather than return to Eritrea. It also addresses a question that has not been discussed at any length in previous studies of the Kassala refugees, namely the family reunion issue. I seek to locate the explanation for third country migration in the global links of the refugees and the hopes and aspirations that these links fuel among the refugees. Given the political conditions in the country of origin and the host country, there is a desire among many of the refugees to resettle in third countries where opportunities for a better life are perceived to be more powerful (Horst, 2003). In the opinion of Crisp it is the desire for greater legal protection that engenders the powerful desire to immigrate to the countries of the north. This is understandable as most African governments confine refugees to restricted sites where they are deprived of freedom of movement and residence rights (Crisp, 2002).

African refugees are furthermore increasingly facing hostility from local populations. Crisp (2004, p. 8), for example, states that the "majority of the refugees in Africa are exposed to more dangers than before due to the different conflicts with the local people". Most likely the opportunity to be naturalised and that human rights are respected also encourages refugees to seek resettlement opportunities (Koser, 2002).

Many of the refugees in Kassala are globally connected with relatives, friends, and former neighbours who live in Western Europe, North America, the Gulf States and Australia. These connections are facilitated through telecommunications and e-mails. For example, many of the Eritrean refugees covered in the study are in constant receipt of information, gifts and remittances. The information about conditions of living in the countries of the north fuels the dream of resettlement even when the possibility is not there (Kibreab, 2005). This is not surprising in view of the great contrast between lives in Sudan and in the resettlement countries.

In spite of the conventional views that refugees are traumatised and weak, refugees should be regarded as rational human beings who are able to undertake cost-benefit analyses of the various possible options that may be available to them (Kibreab, 1993). Not surprisingly therefore most of them make tremendous efforts to re-immigrate to the countries where their living conditions could be greatly improved. Many emigrant countries of the world lack state structures that ensure civil, political and social rights (Horst, 2003). Refugees are expected to return to their countries of origin once the conditions that caused their flight are eliminated. Sometimes, refugees may even return before the conditions that forced them to flee come to an end if the conditions in the countries of asylum are unbearable. Refugees are entitled to naturalise only in a very few countries of the south (Kibreab, 1992). In most countries citizenship rights are enjoyed exclusively by natives of the country. Refugees are excluded from civil, political and social rights. Kibreab, for example, observes:

> "In these countries, even nationals lack the political, social socio-cultural rights they deserve. Refugees in Africa flee to neighbouring countries. In most cases, these countries do not respect human rights issues. The refugees end up in a situation that is no better than the one they left behind. In African refugee producing countries, there is nothing to suggest that the party or

> the front that replaces the old oppressive government, promising freedom and prosperity, would tolerate ideological ethnic and religious differences. If such a party pursues the "winner receives everything" policy, which is unfortunately a common problem in many post conflict societies, not only may the "other" be unable to achieve full citizenship, but also new conflicts may erupt, victimising minority or marginal groups, including those who return to escape from conditions of inequality and marginality in countries of asylum. History is replete with examples in which former freedom fighters have turned into dictators" (Kibreab, 2003, p. 61).

He further observes:

> "The only way to escape such a situation is by migrating to countries that support human rights. This is why many refugees are willing to take great risks in order to live in peace and in a respected environment. Western countries stand for human rights to their nationals and partly to their immigrants. Even though refugees may lose cultural affinity with the host population, they still enjoy their rights to some degree." (ibid., p. 61).

As we shall see in the following section, for some of the study groups resettlement to third countries represents an escape from oppressive conditions and provides an opportunity for family reunion as well as for better life opportunities.

Resettlement: A Rare Opportunity

The dream of resettlement[1] is thus one of the factors that influence the refugees' decision regarding return. This is so in spite of the fact that, most countries in the north have adopted restrictive entrance policies since the mid-1980s. One way through which migrants still can enter these countries is by family reunion. It needs to be borne in mind however that the term family is usually interpreted in a much narrower sense in countries of the north than what generally applies in countries of the south. In most countries in the north, the term family only applies to spouses and minor dependants – children of the family concerned. However, there are variations to the restrictive definitions. In some countries such as Australia, even non-immediate family members can arrange for the resettlement of relatives belonging to the extended family. Many Eritreans have migrated through this channel to Australia (Hassanen, 2000).

Eritrean refugees who live in Western European countries where there are no such resettlement opportunities have to resort to other methods to enable their relatives to join them such as entering the country illegally, overstaying a temporary visa, or applying for asylum and going into hiding after a rejection (Brekke, 2004). Horst (2003a) also found similar practices in her study of Somali refugees in Kenya. She noted, that refugees who already are resettled in a third country help their kin who are left behind by sending money in order to finance their trip. Koser (2000) points out that those refugees who help those who are left behind are the ones that enjoy a secure and stable life in their host country.[2] In most cases the refugees who assist their relatives are not only citizens of their host countries but they also have maintained strong links with their country of origin. That the desire to be resettled in third countries is one of the factors that encourage Eritrean refugees to remain in Sudan can be seen from the following testimony.

> "The whole family is dependent on our daughter's remittances. We receive remittances of one thousand dollars from her every three months through *hawala*. My intention is to be reunited with her. We are one family and we should be together. Being reunited with our daughters would also mean independence because we will be able to earn our own income."[3]

This woman's testimony shows that her ultimate intention is not to return home but to immigrate to Europe to be reunited with her daughters. That many Eritrean refugees in Kassala seem most determined to be resettled in a third country is also demonstrated by the following example.

> "I am waiting for my sister who lives in one of the Scandinavian countries to help me be re-united with her. I already started processing my case. I will feel as if I am born again the day I leave this country and land in one of the Scandinavian countries. Then I will live like a human being, enjoying all the '*Álkirat.*' [4] It will be good for my sister too because she will not need to work as hard as she does now. I will relieve her. What can possibly go wrong in such a modern society where everything is advanced and the people are nice? They are educated. Educated people are nice because they think rationally. You see, the *kawajat*[5] how they speak and how they behave, they are very nice and humane. They even offer citizenship. It is unbelievable! Which country in Africa or which Arab country gives citizenship rights to non-nationals. My sister is a citizen of that country. In Saudi Arabia, my relatives who have lived there for years cannot send their children to university. Can you imagine that in Europe people can get citizenship even before having lived there for ten years! Once I reach there, I will start working and take care of my family. She [my sister] can do whatever she wants to do – marry, study any thing really. She took care of us since she was 20 years old. I will work and take care of the family and then I will do the same with my brother who is working as a computer assistant. We are three sisters and one brother. My father died when my youngest sister was three years old. His death was tragic. He stepped on a landmine, which then exploded. This is how he died. Since then my mum was the one who raised us and my sister took care of financial responsibility."[6]

The testimony of this woman indicates that some of the young people are discontented with their lives in the countries of the north. The imagined view and often unrealistic of conditions in these societies held by many Eritreans and other refugees seems to reinforce their determination to seek resettlement in these countries. This expectation hinders a more realistic understanding of what realities they might have to confront in these host countries. The images that would-be migrants hold are based on information that is disseminated through the networks of transnational communities as well as by the way that Europe and Australia publicize their politics, lifestyles, and welfare systems (Hannerz, 1996). Like the rest of us, refugees are well informed about the world around them. Information is easy to get by means of the new technology and information systems such as Internet, satellite TV etc. However, what is special about the situation of those refugees hoping to resettle in a northern country is that the information they have access to, in many ways correct information, is interpreted selectively. Facts that somehow distort the idealised images and expectations the refugees have are reduced in significance or disregarded (Castles and Miller, 1998).

Do the Refugees in Kassala Enjoy Protection?

By definition, refugees leave their country of origin due to lack of security and safety. If the country of asylum does not offer security and safety, then the refugees have no other option than to search for another place where they can find security and safety. Refugees in developing countries settle in the first country of asylum under harsh restrictive regulations. As pointed out by Horst (2003) these countries are also known for their disregard of human rights. Despite the harsh situation, most refugees still stay in the first countries of asylum for lack of choice. Many of the refugees in Kassala are very aware of and experience these insecurities in Sudan. One respondent expressed this awareness thus:

> "I left Eritrea because it was not safe. There is no security in Sudan either. The situation in Sudan is as dangerous as the situation in Eritrea. The Eritrean government has agents in Sudan. They can do whatever they want. It is all a question of money. Anyone with cash can by bribing the police and the authorities do anything. The Eritrean government has undercover agents all over Sudan. They come to Sudan as civilians from time to time to kill or hijack people from the refugee camps and urban areas. Being a refugee in Sudan does not mean you are safe. The country's security forces are corrupt. Refugees have their own means of protecting themselves. Those who are lucky leave the country and those who are not continue to suffer and die of anxiety. Some commit suicide for fear of being hijacked by PFDJ agents.[7] As for my wife and me, we cannot feel safe until we leave Sudan. Our dream is to be settled in the West where there is freedom and liberty. This is the main reason why I want to be resettled in a third country."[8]

It is confusing that in the group in my case study two opposed positions about return migration seem to be prevalent. On the one hand there is a group that is completely dissatisfied with their stay in Sudan. They are continually seeking ways to migrate to a third country (as the above quoted respondent). On the other hand, another group in the same condition in Sudan is satisfied with their situation in the new country.

Moslem respondents from the Tigre ethnic groups and the Blin ethnic group express unreserved satisfaction with their lives in Sudan and gratitude to the Sudanese people and government for enabling them to stay in Sudan. Some of the respondents even went to the extent of blaming the Eritrean head of the state for his failure to travel to Sudan to express his gratitude to the government and people of Sudan for providing his citizens hospitality. Although there was a time when ELF members were killed in the streets of Kassala, and nobody knows by whom, none of these respondents expressed any fear that Eritrean security agents might sneak into Sudan and cause harm to dissidents or kidnap them across the border. They expressed unreserved confidence in the quality of protection they enjoyed in Sudan.

In contrast to the celebratory remarks of the refugees in Kassala, the respondent quoted above stated that he and his wife were living under a permanent and imminent threat of being assassinated or kidnapped by Eritrean security forces. He also pointed out that the security and police forces in Sudan were so corrupt that it is possible for anyone in possession of cash to eliminate or kidnap refugees. It is in the nature of the problem that the fears that this respondent expresses are hard to substantiate scientifically. However, many people felt this threat and the risks that the respondent points to was common knowledge among my informants.

How could one group of the Eritrean refugees feel so secure and so dignified and the other group feel so insecure and terrified that they believe it is only a question of time before they fall into the merciless grip of the PFDJ agents and Sudanese authorities? There are some possible scenarios that may help us to understand this anomaly. Firstly, those who expressed satisfaction and feeling of being secure are always refugees who fled from Ethiopian occupation during the war of independence; they have remained in Sudan for social, economic and political reasons. Secondly, those who express fear of persecution by Eritrean agents and/or Sudanese authorities are person who have arrived in Sudan after Eritrea's independence, and are those refugees who are actively working against the present system in Eritrea.

Assuming the Eritrean government has agents that cause harm to dissidents in collusion with Sudanese security forces, why don't these refugees fear for their safety? Is it because the Sudanese government provides discriminatory protection or is it because they are not in the wanted list of the Eritrean government? It is unreasonable to argue that the Sudanese government would discriminate between Eritrean refugees. There is no past or present evidence to suggest this to be the case. If the Eritrean government does not cause any harm to the old refugees in spite of its alleged ability to do so, can it be said that the old caseload refugees have well-founded fear of persecution at the hands of the Eritrean government, which did not even, exist as at the time these refugees fled their country. Could it also be that the old refugees who claim that life in exile is more dignifying than life in their country of asylum?

Or could it be due to the fact that the respondent quoted above is exaggerating the dangers dissident groups like himself face in Sudan in order to increase his chances of being resettled to the countries of north or Australia? Although this is a possibility, there is a widespread belief among Eritrean refugees and even among the Eritrean, transnational communities that agents of the Eritrean government sometimes sneak into Sudan either to cause harm or kidnap dissidents. These opposed views about the same reality clearly indicate the difficulty of validating refugee testimonies.

Resettlement as Chain Migration

From the examples given above it is clear that the reasons why the refugees are reluctant to return to Eritrea are many and complex. However, as we shall see in the following analysis, the hope of being resettled in a third country appears to be a major factor that discourages their return to Eritrea. The concern of the Ethiopian government and the Cold War politics of the 1980s seem to have played important roles in the type of people who could emigrate and whether they could be resettled in some host country or not (Zolberg, Suhrke and Aguayo, 1989).

When Eritrea was under Ethiopian rule, most people fled in groups. However, it was difficult for all family members to flee together due to the restrictions imposed by the Ethiopian regime. Therefore, the only way that people could depart from Eritrea was by sending individual family members at different times and through different points of exit. Often families facilitated the departure of youth first because they knew that the regime was opposed to young people leaving the country for fear that they might join the liberation movements. Consequently, more often parents remained behind whilst the young members of their families fled to different parts of the world, mostly to Sudan. By the time parents followed their loved ones to Sudan, some of the latter had already left Sudan either to be resettled in a third country or to go to one of the oil-rich Gulf States. In the 1980s when Ethiopia was under a government that was an ally of the communist countries in the East, the United States government had a resettlement programme for Ethiopian and Eritrean refugees from Sudan. Many Eritreans therefore benefited from the programme and were resettled in the US. The Gulf States were then experiencing an economic boom with a high demand for skilled and unskilled labour. Many Eritreans migrated to the Gulf States in search of employment opportunities (Goutom, 1980).

When parents and children arrived in Sudan, most of them depended on remittances of their children living either in the Gulf States or in northern countries. Over time, those who settled in the West and Australia helped their family members to be re-united with them. Most countries in Western Europe, North America and Australia have policies of family reunification. Under the policy of family reunification large numbers of Eritrean refugees have immigrated to different countries. What is interesting about these family-reunification based emigrations it that new kinships are being re-created in countries such as Australia where there are heavy concentrations of particular ethnic or kinship groups. For example, in Melbourne there is a high concentration of members of the Moslem Blin ethnic group and Tigre-speaking groups from Ali Gidir in Gash Setit (Hassanen, 2001).

The chain of movements and the transnational links

are made possible through the revolution in technology – telecommunications, Internet and air transportation. The reason why there are more family-reunification resettlements in Australia and North America than in Western Europe is because of more liberal reunification schemes, where the cost of transportation for reunification is sometimes funded by the receiving countries (Australia, Canada and the US), which is not the case in Europe. In Australia and North America, the definition of a family also seems to be broader. It is worth mentioning in connection to this that the hope of being resettled in the countries of north and Australia is not only discouraging people from returning to Eritrea under different pretexts, including fear of persecution, the desire to be resettled in one of the countries of north has also been encouraging people to use the most dangerous passageways (Pateman, 2001; Morris and Crosland, 2001; Mavris, 2002).

In the rapidly globalising world, the factors that motivate people to leave their homes and to return to their homes have become extremely complex. One respondent in Kassala, for example, reported:

> "I have a sister and a brother who have been living in Norway for about ten years. My sister is married and has her own family. My brother is a student. I am going to join them soon. My mother is in Eritrea. I am waiting for her because she will come to say goodbye to me. She will arrive anytime. I will leave Sudan after I say good-bye to my mother. I am migrating to Norway because I have my siblings there who can take care of me. If I got a chance of being resettled elsewhere, I would not have taken it because before I can accept such an offer I need to know about the place and the people where I would be going. I would only go to a country where I have someone who could help me through the first phase of my adaptation. Otherwise, it would be difficult for me to travel to a place where I have never been to and I do not know anything about. As a girl, it would be very difficult for me to travel alone to a place where I do not have anybody. What I

know about Norway is not that much. However, I know the climate is very cold and during the winter season, there is snow everywhere. I think I will be fine. I miss my siblings so much. As long as I have them, nothing can go wrong. I believe in God, and everything that happens to me is in the hands of God. Anyway, all is in God's hands. I can only say this much about my future and the rest is up to God."[9]

For this respondent and many others, the countries that are most attractive are the ones where their families already live. This type of emigration is motivated by the need to join loved ones and more importantly by the need to improve one's life chances through education, employment and for the aged through receipt of income or support of family members.

This particular respondent is not only looking for any resettlement opportunity but rather for a country where she has siblings who could provide her with the necessary support. Her aim is not only to be reunited with her sister and brother; she also has other visions for her future. However, all her visions are attached to the opportunity that was created by the family reunion in Norway.

The data elicited from many interviewees show that those who intend to migrate to the countries of north and Australia have unrealistic expectations about life in these countries. What they know about these living conditions is what they pick up from images portrayed through the media, in advertising and commercials, and hearsay from relatives and friends who report back home. Those who wish to migrate see the positive things they hope to access, such as getting a higher education, being respected and appreciated by local people on equal terms with the rest of the population etc. They tend to disregard information that reports of the hardships that immigrants generally encounter. There is practically no awareness regarding how hard life can turn out to be. It is common knowledge that emigrant communities in spite of the good pictures they come with, find the adjustment process daunting and sometimes depressing (Westin, 1999). This is because there is always a mismatch between expectations and realities on the ground.

Most respondents I interviewed, hope that upon arrival in the resettlement countries, they will be able to study at universities, become engineers, medical doctors, scientists, nurses, teachers, etc. After graduation, they assume they would face no difficulties in finding jobs commensurate with their newly acquired professions. However, the reality on the ground is often very different. For instance, in Sweden and elsewhere in Western Europe there are many highly qualified professionals among immigrant communities working as taxi-, bus- and train drivers and ticket sellers. Migration studies point out that migrants leave their country of origin to seek a better life elsewhere. The reality at the destination is more complicated. Some make it, others don't, but those who have high aspirations tend to completely discount the potential difficulties. Not all migrants are having a better life than they had before. For some migration is becoming a key to success while for others it is not (Westin, 1999).

Different life experiences in the past and the present shape the way refugees live and think in their new destinations. In the West, refugees are struggling against many odds such as marginalisation and discrimination. They may be formally and legally included in the host societies, but in reality, they may be excluded socially.

Conclusion

As pointed out above there is one category of refugees who want to be resettled to other countries. Why would they rather migrate to a third country than return to Eritrea? Once again, as important as the economic factors are to explain their motives, an additional factor is political. The refugees make this choice, first of all because they do not feel that their country of origin can give them what they want, that is to live in a democratic country. Moreover, they also want to live with their relatives who are settled in third countries where the general standard of living is much higher than in Sudan and Eritrea – or at least they imagine it to be higher. This is because the refugees have an image of what life in the countries of north can be. These images are based on hearsay, stories and rumours disseminated through the channels of transnational communities, but also through the ways in which European countries, the United States, Canada and Australia present their politics, lifestyles, and welfare systems. As indicated in this chapter, refugees are well informed about the world around them, made accessible to them through the modern technology as well as modern information systems such as internet, satellite TV etc. Their knowledge of the favourable conditions in countries of north and Australia contributes to their wish for a third country migration. However, the wish of the refugees to leave first Eritrea and later Sudan is above all because of the abusive regimes in those countries.

Notes:

1. Although, there is a right to claim asylum, there is no right to claim resettlement. This is despite resettlement to third country is one of the three durable solutions to refugee problem. Moreover, it is important to point out that resettlement countries determine the standard by which refugees ask for resettlement and the individuals are considered for resettlement according to the norm of UNHCR. According to an interview with one of the UNHCR staff in Khartoum (2002) and from what I observed it seems that the selection of the refugees for resettlement is more connected to needs and priorities of the resettlement countries than to the needs and rights of the refugees. For instance, most of the refugees that pass the interview test are most of the time healthy, young, educated etc.

2. People who practice their rights regarding stay permit are those who have some kind of financial income

3. A mother, head of the household who left Eritrea in 1975. She is illiterate and lives on remittances. She is a Moslem and belongs to the Blin clan. The interview took place in the informant's house and was held in Tigre. It was translated into English by me, Kassala, 2002.

4. Good things.

5. White people.

6. Female, single, left Eritrea in the beginning as a teenager, high school graduate, Moslem, belonging to the Tigre clan and sympathises with the ELF. The interview took place in Kassala market in a Photoshop and was held in Arabic. The interview was translated into by English, Kassala, 2002.

7. The ruling political party in Eritrea.

8. Family man who, left Eritrea in 1999. He is a high school graduate and a former high-ranking military officer of PFDJ. He is a Christian, and belongs to the Tigrinya clan. For security reasons the respondent resides in Muraba'at and sometimes in Abukamsa, The interview was held in Tigrinya and translated into English by me, Kassala, 2002.

9. Female who left Eritrea in 1979. She is single, and a university graduate She is a Moslem, belongs to the Tigre clan and is a member of ELF. She resides in Muraba'at. The interview also took place in Muraba'at in the UNHCR RSD registration centre, and was conducted in Tigre. It was translated into English by me, Kassala, 2002.

7

HOST-COUNTRY RELATED ISSUES

Survival Strategies, Livelihoods and Integration

The aim of this chapter is to analyse the host country-related factors that influence the decision making regarding return migration by examining the integration of the refugees in Kassala town and analysing the survival strategies of the refugees and their livelihood systems. Integration does have an impact on the way the refugees view life in their host country as well as their country of origin.

As pointed out earlier, the integration of Eritrean refugees in Sudan has been discussed by researchers such as Kibreab (1996), Bulcha (1988) and Kuhlman (1994), who have carried out studies in eastern Sudan, but these studies have not focussed on the relationship between religion and integration, which is an issue this study addresses.

According to the policy of the government of Sudan, refugees are required to settle in government-designated areas. However, many Eritrean refugees settle outside such areas, thus defying the government policy. Formally, no Eritrean refugee is allowed to settle in urban areas, but there are many who do so in Kassala (Karadawi, 1999). The large majority of the refugees that left Eritrea before independence covered in this study, especially those who are from the Moslem Tigre and Blin clans, have informally regularised their stay in Kassala through various means. Many informants gave me information about how this could be done.

However, those who settle in urban areas are excluded from receiving any form of international assistance. To survive, they have to manage economically on their own. In such a situation the need for wide social networks is an imperative necessity and it is through these that most of the households make ends meet. In the absence of such wide social networks, most of the families would be vulnerable financially and socially. Except for households who depend on remittances from relatives, the livelihoods of the rest of the refugees are to some extent a function of the level of integration. It may therefore be interesting to ask the following questions: To what extent and in what ways are the refugees living in Kassala town integrated into the host society? What are the factors that influence integration? How do the refugee families make ends meet? What are their survival strategies? Integration is about social interaction of peoples from different ethnic, social or economic backgrounds. It is about mutual acceptance and cooperation (Kuhlman, 1991).

Factors that Influence Integration

Eritrean refugees in the town of Kassala live among the host population and therefore, at least theoretically, they have an opportunity to interact with these people. This interaction does not take place in a vacuum. To a certain extent, it requires some degree of socialisation and a platform where such a socialisation process occurs. A reasonable assumption is that integration is only possible in a situation in which the two groups concerned, in this case Eritrean refugees and their hosts, are settled in the same neighbourhood, thus enabling participation in various domains of everyday life (Gordon, 1964; Dias, 1993).

For integration and social interaction to take place, physical proximity is a necessary condition (Kuhlman, 1991). My study confirms that the places where social interactions take place include neighbourhoods, markets, mosques and schools. This means the significance of the mosque is not limited to acts of praying, mourning, performance of religious rituals, celebration of holy days, but it also functions as a social platform where refugees and locals get a chance to know each other and to exchange experiences. Since both the refugees and the local inhabitants share common religion – Islam – this provides an opportunity for social interaction and exchange of ideas and experiences. Common religious values tend to create a foundation for mutual interaction (Kuhlman, 1994).

The data of this study indicate that Islam inculcates certain values and norms on believers. One of the central core values of Islam is the duty to extend solidarity to fellow Moslems who are in need. Those who are displaced by forces beyond their control are considered as people in need. Islam instructs Moslems on how they should live their lives, as well as how they should share their material possessions, how they should dress, pray, eat, pay taxes, etc. Even though, it would be unreasonable to say that all Moslems observe all these duties, the central aim of these teachings and instructions is to create solidarity within the Moslem community or nation – the *umma* – worldwide. The fact that the large majority of the refugees and the local hosts are Moslems means that both adhere to the same core values that guide their lives. Hence, Islam is the glue that ties the Moslem refugees and the local communities together. It is important to mention that the majority of the Eritrean refugees in Kassala do not only share religious values with the local people, but also share common culture, such as language, marriage ceremonies and religious celebrations and rituals. Therefore, the ethnic groups that have high potential to become integrated in the host country are those who have affinity with the Sudanese, such as the Tigre speaking group (Bayoumi, 1999).

The longer people live in one area, the more they adapt to the particular way of life prevalent in that area. The more refugees adapt to the conditions in the country of asylum, in the context of a welcoming host population and a government policy that is indirectly helping, the more they are likely to question return migration as a viable option (Gordon, 1964). When one talks to such refugees about whether they intend to return or not, they express no doubt about the fact that they will remain Eritreans.

As my study tells me, what is interesting about the refugees who to some degree have found a new home in Kassala is that the status of the large majority has not been formally regularised. It was evident from my fieldwork that locals generally accept those Eritreans with whom they share common religion and ethnicity, regardless of people's social status. As long as the relationship between the Eritrean government and Sudanese government remains hostile and consequently both continue to use each other's citizens to undermine each other, this strategy may continue to work. The question is whether the refugees can count on this when it is common knowledge that the two governments' moods are constantly swinging. Notwithstanding the potential dangers, at present there is a cordial relationship between the local populations and the refugees who share common religion, ethnicity, language and way of life. These refugees are informally integrated into the local communities. However, this is only true among those who share common religion, ethnicity, and language with the local population. Eritrean refugees who do not share such commonalities with the local population have fewer opportunities to interact in everyday life with the locals and to participate in joint arrangements of various kinds. In this sense one could say that they are not really integrated in to the local community.

One of the factors that facilitate the integration of refugees in Kassala town is obviously a welcoming local population. Although the government is adverse to the presence of refugees in the town, through informal cooperation with the local population, the refugees avoid being harassed by it. As pointed out earlier, Islam is a religion with a lot of rules and regulations. It creates a bond between believers in different societies. Notwithstanding the differences of their languages, culture or geography, Moslems find a common platform in places, such as mosques etc.

Sources of Livelihood and Survival Strategies

In the studies of forced migration, the role of religion in integration and livelihood systems of refugees in first countries of asylum has been neglected. The informants I have met and spoke to about these matters acknowledge the critical role religion plays for their integration and livelihoods o in Kassala town. As noted earlier, the refugees that are well integrated into the host society are the ones who share Islamic values with the local people.

Three socio-economic categories are identified among the refugees in Kassala town. The first category consists of those who enjoy strong social networks with members of the local population. These strong links and social networks enable them to obtain business licences, to buy property, to access local schools, healthcare and for their children to take part in Sudanese national examinations. These are the refugees in the first category who over time managed to use their social networks and political connections with the regional authorities to obtain proper documentation that enable them to pass as Sudanese.

The second group comprises refugees who do not share common ethnicity and religion with the local population. The groups that belong to this category include the Tigrinya and Blin Christians. Although, these groups are disadvantaged because they lack the necessary social networks their children attend Sudanese schools and they have access to public health care. However, because of the ethnic and religious boundaries vis-à-vis the local population they are less integrated. Their networks do not link up with the networks of the local population. The third group does not have any distinct ethnic or religious identity. In this category fall people whose livelihoods are dependent on remittances sent by relatives living in the Gulf States, Western Europe, North America and Australia.

Social networks are the key to survival for these refugees. Without such links, it is impossible for any refugee to engage in an occupation, or to trade in the formal and informal sectors of the regional and national economy. More importantly, whether a refugee is able to obtain *Jinsia* depends very much on one's access to social networks that are based as we saw earlier, on common religion and ethnicity. Without *Jinsia* documents, the refugees cannot move between places in search of employment or trading opportunities or in pursuit of social activities (Kibreab, 1996; Kuhlman, 1994).

A closer look at the refugees' survival strategies shows that most of them are innovative and make ends meet by engaging in diverse income-generating activities in spite of the adverse conditions that characterise their lives. When refugees lose everything they possessed and are faced with new but familiar problems, they are forced either to rethink the things which they took for granted previously or to engage in activities which they previously avoided (Edward, 2000). When the refugees are faced with a life and death situation, they free their energy to earn an income in order to prevent the danger of subsistence insecurity. In spite of the hardships and suffering that accompany involuntary displacement; many refugee communities transform these adverse conditions into strength (Kibreab, 2003). The findings of this study confirm Kibreab's observation.

Those who have the greatest success are the refugees who have connections with government officials at various levels, either through ethnic or faith based social networks. Many of the refugees who share common identity with some of the local population regularly borrow ID cards of some of the local population to obtain a business licence. In order to facilitate their possibility to earn an income or to move between places, those refugees who have not regularised their stay generally have to resort to other kinds of means to engage in income-generating activities such as starting a business. It is interesting to note that it is in the process of undertaking such activities that the refugees manage to build their ties with the local economy and society. Over time, the social interactions become consolidated enabling the refugees who share common ethnicity and religion with the locals to become part of the local communities.

For some refugees, cross-border trade represents a profitable source of income. This is demonstrated by the following example, which by no means is isolated.

Ahmad was born in Eritrea. He fled with his family and came to Sudan at the age of six. His family's displacement was caused by Ethiopian persecution. Ahmad is a Tigre-speaker who fled with his mother, father and his five siblings. He attended a Sudanese school but dropped out in year nine. He engaged in different income-generating activities, such as goldsmith apprenticeship and shop keeping. Twenty years ago, he began trading between Kassala, Port Sudan and Eritrea. At the time of the interview, Ahmad was a wealthy man with a family of his own. He owned property and other forms of wealth. He accumulated a considerable amount of wealth because his cross-border business activities, i.e. between Eritrea and Sudan, were extremely profitable. In an interview, he said:

> "I have been in this business for about twenty years. Doing business in Sudan is not that difficult. It only requires strong ties with the authorities and the local people. As a merchant, one has to know how to handle difficulties with the authorities. The market is booming and the demand for foodstuffs, spices, coffee, etc. [imported from Eritrea] is very high in Kassala. In order to avoid problems with the authorities, one should be ready to pay bribes. This is even true for Sudanese businessmen. Bribing is a common practice in this country. No business transaction can take place without payment of bribes."[1]

This respondent, like many others in his situation, runs businesses in Kassala town. The factors that facilitate the business activities of refugees such as Ahmad include common language, religion and ethnicity. Social networks especially the ones that link refugees with government officials are essential in the survival strategies of the refugee families. There are even some individual refuges that make a living by mediating between refugees and officials in government, the police or security. For those who lack direct links with police or security officers and government officials, the presence of a go-between or middleman is indispensable. This is because in spite of the fact that payments of bribes are widespread and part of the refugees' everyday life, it is still illegal. If a refugee is caught bribing a government official, his safety may be put in danger unless he is able to bribe those who catch him or her in the act of committing the offence of bribe payment.

Those in the first category who share common ethnicity and religion with the locals often have easy access to people in power without any problem. However, this does not provide adequate protection against any kind of abuse because those who help would never dare to come in the open to defend the refugees when he needs them the most. The refugees who have common ethnicity, religion and language with the locals are less vulnerable than others because they can more easily pass as Sudanese.

Even though Ahmad as a non-Sudanese national did not enjoy formal citizenship rights, his wealth and his contacts contributed to his further success in his business activities. Although people such as Ahmad do not openly admit that they possess informal Sudanese *Jinsia*, there is no doubt that he could have never reached his present economic status without one. He is one of many Eritrean refugees who have found new places in the country of asylum by constructing a new and secure source of livelihood. This life, no matter how successful, is not however without risks. It is not only poor refugees who are vulnerable to abuse at the hands of police officers and government officials. Even successful refugees are also constantly harassed by bribe-hungry people in authority. The refugees who make it have the protection of patrons but it is a costly enterprise. Failure to have a patron usually means that one is vulnerable to endless problems.

As this study indicates in terms of economic status, the conditions of the refugees are not worse than those of the poor local Sudanese. In fact, some refugees have better housing conditions than some of the poor local population. This is true in the three areas covered in the study. Poverty does not seem to be a function of legal status. Being a refugee or a national does not seem to determine the economic situation of the people in the study areas. There are local people as well as refugees who live in poor conditions. Areas housing the poor are concentrated to the periphery of the town. It was apparent to me in my fieldwork that most of the local people who live in the poor areas belong to the Beja ethnic groups – the most marginalised ethnic group in Eastern Sudan. Ironically, these are the indigenous people of the area.

Although those refugees who share common religion, ethnicity, language and way of life with the host population are more integrated than the other refugee groups, there is no discernible difference in the attitudes towards their country of origin between the well-integrated and less well-integrated groups. All express loyalty and strong sense of belonging to their country of origin and view the local people as kind and generous. Among the populations covered in the study, the Tigre speakers are the majority. The Tigre speakers in spite of their common language belong to different ethnic groups which prior to their arrival in Sudan had nothing or little to do with each other. Among the Tigre-speaking refugees, the Beni Amer represents the overwhelming majority. The next largest Tigre-speaking groups are from the Senhit (now Anseba) region. A considerable proportion of the Maria has returned to Eritrea in response to the political changes that took place in their country of origin (Kibreab, 2002).

The degree of social integration and quality of reception afforded by the local population appears to be a function of ethnicity and religion. The case of the Moslem Blin is anomalous in this regard. Notwithstanding the fact that they lack any form of ethnic association with the local Sudanese in Kassala town, they enjoy an equal degree of integration as the Tigre speaking clans. This not true of the Christian Blin.

Figure 1. Houses of families who live on remittances. The quality of the house according to the local standard is the highest in Kassala. Sadia Hassanen took the photo in 1996.

It is interesting to see why the Moslem Blin is treated differently from the rest. In terms of ethnicity, the Moslem Blin has nothing in common with the local Sudanese. Their integration is a function of their religion or perhaps their political stance. In fact notwithstanding the fact that they came to Sudan as refugees and there are no Sudanese that share their ethnic identity, the government of Sudan has allowed them to establish their own *Omedia* – the lowest unit in the structure of local administration based on common tribal affiliation.

It is interesting why this is taking place, formally, the Sudanese local or national government does not want to regularise the stay of the refugees in the urban areas. However, most of the Blin Moslems and the Beni Amer have acquired Sudanese *Jinsia* (nationality) without the official acknowledgement of the government.

Nevertheless, all the respondents are grateful to the government of Sudan and appreciate the local people's hospitality. When asked why they settled in Kassala rather than elsewhere, they mentioned the following reasons: (i) Some had relatives in Kassala who provided support and information at the initial stage; (ii) The ELF had its office in the town and this engendered some sense of security and familiarity; (iii) Kassala is close to the Eritrean border; (iv) A substantial proportion of the local people speak Tigre. It made sense for the Tigre-speaking refugees to settle in Kassala town; (v) Common ethnicity with a substantial proportion of the inhabitants of Kassala and the refugees are from the Beja and this was a major factor that influenced the choice of destination.

Refugees Without Social Networks

Life for those refugees in the second category, namely, those who are not connected with the local population through ethnicity, common language and in some cases religion, is not easy. The refugees in this group are on the margin rather than part of on mainstream society. To get by, they need to work harder and they face greater obstacles than their compatriots who belong to the well-cushioned members of the Tigre speaking community. However, in spite of these difficulties they face, they are able to earn a living by engaging in diverse income-generating activities. Notwithstanding the fact that they engage in diverse income-generating activities and have been living in Sudan for several decades, they are less integrated into the local community in terms of housing and social networks. Despite these difficulties they are still unwilling to return to Eritrea. According to the 2002 UNHCR memo, most of the refugees that belonged to this category have returned to Eritrea. Their lack of a common cultural heritage with the local people affects their way of living in Kassala and this puts them in a weaker position than those identified earlier in the first category.

In Eastern Sudan in general and in Kassala town in particular, there are very few Christians, which was pointed out to me in informal discussion by representatives of the Red Cross and Red Crescent in Kassala in 2002. The overwhelming majority of the population is Moslem. Christian refugees are disadvantaged because they do not participate in the host population's social life as the first category of Moslem refugees. There are no common platforms for social interaction, e.g. religious rituals and ceremonies. Although there are many refugee families in the first category (Moslems) who own property, I did not come across any family that belongs to the second category (Christians) that owned property in Kassala. It is not clear whether this is due to their religion, i.e. they lack the necessary connections and therefore would not be able to buy houses or whether it is because they do not see themselves as part of local society.

Perhaps they feel marginalised due to their religion and language. The majority of the Christians speak Tigrinya, a Semitic language that belongs to the same family as Arabic, Tigre and Hebrew, but it is not spoken by any group in Sudan. They are also excluded from the host society because they constitute a distinct entity identifiable by their clothing, names and perhaps by their knowledge of Arabic. However, not all Eritrean Moslems speak Arabic, but at least they share similar names and clothing with the local people and this seems to provide them with some sense of protection against abuse and alienation. It is difficult though to say that all the Tigrinya speakers and Christians experience exclusion and alienation as the testimony of following respondent shows.

Celine is a Christian and speaks Tigrinya. She has been living with her husband in Kassala since 1999. Earlier they lived in Wedsherifey refugee camp outside of Kassala town. Celine's husband found a job as a truck driver. As a long distance truck driver, he sometimes he stayed away from home. His employer, a Sudanese, offered him a room in the employer's compound where he could rest when he stayed in Kassala. After some time Celine moved to Kassala to live with her husband. They lived in rented accommodation. At the time of the interview Celine had three children between 4 and 12 years. Celine is positive about the local people. She thought that they were nice and generous, but missed a lot of things that were not available in the neighbourhood. Her relationship with her neighbours and with the family of her husband's boss was friendly. Sometimes she went to the boss's family to help in the kitchen. Her relationship with the local people and the Moslem Eritrean refugees was also friendly. However, she thought that it was due to her own efforts that her relationship with the neighbours and fellow Eritrean Moslem refugees was good. Neither the Sudanese neighbours nor the Moslem Eritrean families made any efforts to nurture a friendly relationship with her. She said that she visited her neighbours and her Moslem compatriots during *Id*, which is an Islamic holiday, and other festivities, but they never reciprocated by returning her visits during Christmas or Easter.[2]

Celine's case is unusual. There were very few of my respondents in her position. Although she has a relatively friendly relationship with her Sudanese neighbours and Eritrean Moslem refugees, as a member of the Christian minority group, she took the initiative to cultivate relationships with her neighbours and with Eritrean refugees who felt at home more than herself because of their shared ethnicity and religion with the local population.

Remittance-based Livelihoods

Remittances play a key role in the livelihoods of many refugee communities. For example, Cindy Horst (2003) found in her study of Somali refugees that remittances play a key role in the livelihoods of many refugee households. However, whether members of a given transnational community send remittances or not is to a large extent dependent on the degree of commitment senders have to receivers. Among Eritrean refugees it is common to feel a strong sense of duty and responsibility to help relatives in need. This constitutes one of the most cherished core values of Eritrean society, which was brought home to me in long discussions with several different families who live on remittances. According to informal conversations with exiled Eritreans in both Stockholm (October, 2003) and Melbourne (August, 2006) these traditional values are continuously renegotiated and are changing, in most cases members of the transnational communities have a strong sense of commitment and duty to help their relatives at home and those stranded in first countries of asylum, e.g. Sudan.

It is important to recognise, however, that the level of commitment of members of the Eritrean transnational communities to their relatives in need is dependent on the strength of ties that link the sender and receiver of remittance. The norm in the Eritrean society as expressed by informants in Sudan as well as in Stockholm and Melbourne, is that parents should be supported by their offspring whenever they are in need or during their old age. An offspring who fails to live up to this expectation is condemned as being unworthy, and depending on the circumstances can suffer loss of reputation. Therefore, many Eritrean emigrants (refugees) see it as a duty to help parents and siblings. Sometimes, members of the Diaspora, especially women go out of their way to help relatives in need. It needs to be pointed out, however, that there are some members of the Diaspora who neglect this duty but they are very few (observation during the fieldwork and my own experience).

It is not uncommon for younger brothers or sisters to send remittances to their older brothers or sisters. In most cases, it is the successful family members who take the responsibility. The responsibility to siblings and cousins may differ. Duty towards close family members is considered more serious. For instance, those who support parents and siblings regularly have to take their commitment very seriously while those who support extended families do not need to be as committed as the former ones. Studies show that such duties are common in most economically poor countries that lack welfare systems (Van Hear, 2003). Children are used as an asset for the future life of the parents. This means taking care of aged parents. In such societies, children are seen as one form of life insurance.

Those who live by remittances receive cash and gifts from members of their extended family. Those respondents who have close relatives such as parents, spouses, or siblings receive regular remittances. However, those who have more distant relatives like cousins or nieces mostly receive money if they ask for it.

In Kassala, the refugees whose livelihoods are dependent on remittances generally have a higher economic status than the majority of the refugee families who rely on incomes derived from their own income-generating activities. This is due to the fact that those who receive remittances have more money to spend than others. Of course, the most successful businessmen, who own property and shops, live in houses that are built of brick or cement with containers of water and access to electricity etc. Those rather few refugees who can afford to live in such houses are those who depend on remittances from close relatives who work abroad, either in the oil rich Arab countries or in the countries of the north.

Among the Moslem refugees who do not have ID cards, it is common to ask for help from their countrymen or neighbours to get by. The testimony of the following key informant is typical of the families who live on remittances.

Shamin is a Moslem, housewife and is 58 years old. She is the mother of eight daughters. She depends on remittances sent by her daughters who live in England. She lives in her own house with three daughters and a grandson. Two of her daughters live in England with their own families. Three others live in Saudi Arabia. The latter are also married. Shamin's house in Kassala is bought by her daughter who lives in England. Her house is built of bricks and consists of five rooms. Shamin left Eritrea in 1975. Her husband lives in Eritrea. She visits him twice a year.[3]

Refugees are not allowed to own property in Sudan (Kibreab, 1987). The reason for this is that Sudanese governments have consistently regarded refugees as temporary guests who should return whenever the conditions that forced them to flee come to an end. Ownership of property indicates some degree of permanency. That is why the government's policy clearly states that refugees should live in temporary accommodation until the conditions that prompted their displacement are eliminated. In spite of this prohibition many of the Eritrean refugees who live by remittances and those who belong to the first category (Moslems) own property.

It is also interesting to note that although refugees own property and are settled in Kassala for more than three decades they still present themselves as refugees and they express the desire to return to Eritrea if there is a change of regime. Whatever plans of return that these refugees may express, looking at their settlement and cultural affinity with the local people it seems fairly obvious that they are settled more or less permanently in Kassala. Their talk of return to Eritrea may never come about, but should rather be understood as a way of expressing one's identity, similar to the traditional Jewish parting phrase: "Next year in Jerusalem".

The examples of people who live by remittances and who run successful business activities should not be construed to imply that all the refugees in Kassala are well off.

The following example may demonstrate this. Naser is a Moslem who comes from the border area of Eritrea. He is married and has seven children. He works in the vegetable market in Kassala. His children attend Sudanese schools. The family lives in Muraba'at on the outskirts of Kassala in a hut constructed of thatch grass and mud. The family has no access to water or electricity. Naser and his family live from hand to mouth. He says that for his clan, the Tigre, it is easy to access local services because they are considered as members of the local population. Notwithstanding the fact that this respondent comes from a group that is almost fully integrated into the local communities, his family remains very poor.

These two examples show that the socio-economic status of the families even among those who belong to the first category of refugees that are well connected to the local communities is differentiated. Some are well off, e.g. Ahmad and his like while others such as Naser depend for their survival on a daily income. The differences in their socio-economic status notwithstanding, nearly all the refugees who belong to th e first category are connected to the local society and economy.

Neither of these two respondents, like many others in their communities, has ever applied for refugee status. It is questionable therefore whether the label refugee applies to them. This is in spite of the fact that they fled their country due to Ethiopian persecution. It is not only refugees belonging to this category who did not seek asylum. The large majority did not apply for asylum. Since they come from a country where there was war and disorder, it was assumed by the government authorities and UNHCR that they are prima facie refugees even though their status was never determined.

Among the families that live by remittances, female-headed households are over-represented. The phenomenon of an absent "father" is common among these families. The reasons for this differ from family to family. Some of the 'fathers' never left Eritrea while others immigrated to the Gulf-States.

Dina is a 32 years old single female who lives with her family in Kassala. She said,

> "Since we arrived in Kassala, my brother has supported us. My family and I came to Sudan in 1975 from a town called Agordat in Eritrea. Since then, we have been dependent on my brother, who is living and working in Saudi Arabia. I have also two sisters who live in Europe and we get some money from them too, it is my brother who sends remittances on a monthly basis."[4]

Displacement has broken traditional practices of remittance sending and this has increasingly become a function of capability more than anything else. In fact in most cases, it is now women rather than men who bear the responsibility of sending remittances to members of their families at home or in first countries of asylum (Lambert, Briere, Sadoulet, Janvry and Lambert, 2002). The changes are mainly due to the fact more women are increasingly becoming either joint or sole breadwinners. Whether it is true or not it is also commonly believed among Eritreans these days that women are more empathetic to the plight of their family members than men and that may explain the reason why women tend to remit more cash to their relatives than their male counter parts (group discussion with families in Kassala).

The Effect of Remittances on Return

The refugees in Kassala are interconnected globally. Being globally connected through relatives entails many direct and indirect consequences. One of the consequences is the obsession or the powerful dream of resettlement to one of the countries in the north. This powerful desire of resettlement is to a large extent precipitated by the 'misleading' signals sent by those refugees who are resettled in these countries of the north. Remittance sending is not only about sending cash, it is also about sending signals about favourable conditions in those countries where the remittance sender lives. The money sending is connected to the sender having a better and a happier life. The recipients assume that unless the sender concerned has excess cash, s/he would not send so much money. Recipients have no clue about how the cash they receive from their relatives is earned.

Figure 2. Example of how families with poor economy are settled in Kassala. The photo was taken by Sadia Hassanen in 1996.

Most refugees in the Diaspora do low-paying and menial jobs to earn their living. It is from those meagre earnings they send remittances to their relatives at home or to those who are 'stranded' in first countries of asylum. The fact that they send a lot of money does not mean that they have extra cash (Akuei, 2005). Most of them forgo their medium and long-term family interests to help their relatives in need. The social obligation of helping relatives in need is so burdensome that there are many families that breakdown because of such pressure (ibid.). Those who live on remittances from relatives abroad are not aware of the conditions under which the cash they receive is earned and the tension they cause to families by those at the sending end. The obsession is far more powerful among the youth. Many of the latter take drastic actions such as risking their lives in order to settle in a third country (Discussion with families and youth in Kassala).

Remittances also create dependency. In Kassala some families are completely dependent on the money that is sent to them from relatives abroad while others are semi-dependent. Those who have been dependent on remittances for many years without themselves contributing to their subsistence, waste their time dreaming about being resettled which they may never realise in their lifetime. This fantasy was common among the refugees who had members of their families in a developed country. In a group discussion, I asked young boys whether they would consider staying in Kassala if they found suitable jobs after their graduations. They all said 'No!' When asked to state the reasons, they all agreed that the jobs that are available in Kassala do not require skills, they are low paid and boring. Their dream was to be resettled in one of the rich countries in the north.

The fact that they thought resettlement would mean doing professional jobs is an indication of misinformation. They are not aware of the fact that those relatives who are sending remittances to sustain them are doing the most monotonous and unskilled jobs wherever they are. They would have probably been put off if they learned that the money they were living on was earned from washing of toilets and dishes and cleaning. This may to some extent also be due to the fault of those who send remittances. It is improbable that that they inform their relatives about the kind of work they are doing to earn a living.

None of those who lived by remittances in Kassala wanted to know about the difficulties refugees and immigrants face in Europe and elsewhere. For instance, most immigrants (refugees) in Sweden work within the sector that requires no skills, such as cleaning homes for the elderly or dishwashing in restaurants. The respondents were neither willing nor ready to hear what I told them about the conditions under which their relatives earned the cash they sent to them.

The cause of the misinformation could be the remittance senders' reluctance to reveal their occupation. Their relatives believe that they are working as qualified professionals. Remittance senders may withhold information from their relatives either because they feel some sense of embarrassment or not to dishearten their relatives. If one worked as a teacher prior to one's departure from home and now is working as a cleaner in one of the countries of the north, it is contrary to expectation that such a person would share such sad information with his or her relatives at home or in first countries of asylum. This is not unique to Eritrean communities. Horst (2003) found in her study on Somalis living in Holland that if they told their relatives about the difficulties they faced in Holland, they would be accused of telling lies in order to avoid the responsibility of sending remittances to relatives in need.

However, a few respondents seemed to be aware of the difficulties their relatives in the rich countries might face as the result of the duty of helping relatives. This is implied in the following testimony.

> "My siblings live in one of the Scandinavian countries. Financially we are dependent on them; they send us money from time to time. My husband has casual jobs, and his salary is not enough, therefore I am mostly

> dependent on my sisters' help. I know my sisters have their own problems; they are married and have children. Despite that they call me every two or three weeks to know how I am doing. I never ask them to help me but they know my situation from before and as soon as they get the chance, they send me some money."[5]

The responsibility of helping relatives is not limited to immediate family members. In most parts of Eritrea, the extended family is a norm and this tends to increase the burden of remittance senders who are expected to help distant relatives or cousins. The following example demonstrates this.

> "I am the only daughter, thus, my parents got me married at an early age, in order to be close to them here in Kassala. My father works as a guard, and my mother is a homemaker. My husband is a truck driver. He supports the whole family but his salary is not enough. I have relatives (cousins) in Australia. They send us money every now and then. Sometimes they also send us clothes and perfumes."[6]

The pressure on remittance senders is not limited to supporting close relatives financially but they are also expected to facilitate the resettlement of their relatives either legally or illegally. Not only is the cost of helping relatives to enter the rich countries in the north very high but also the risk of failure after payment of prohibitive amounts is very high. Those at the receiving end either discount the pressure on their relatives abroad or they are not aware of it. They are single-mindedly focused on being resettled no matter what. In this context, the idea of return to Eritrea does not figure out in the options the refugees consider desirable. However, when you ask any refugee why s/he is not returning to Eritrea, they would never say, 'we don't want to return, because our dream is to be resettled in one of the rich countries.' Instead they would say, 'There are human rights violations, there is no freedom, no democracy, the government oppresses Moslems, etc.' Some of these arguments may not necessarily be wrong, but they either have little or nothing to do with the reasons why those who are desperate to be resettled have not returned to Eritrea.

Kibreab (1996b) noted that the decision-making among the refugee communities in Sudan is based on the livelihood considerations upon return. Then the question is: Why do refugees who live on remittances prefer to stay in Sudan when they are dependent on outside economic support? Moreover, according to the refugees who returned to Eritrea but later came back to Sudan, the cost of living in Sudan is much higher than in Eritrea.

CONCLUSION

The refugees in Sudan are not allowed to settle in urban areas. However, as noted in this chapter some of them are settled in an urban area and integrated in the social life and economy of the host country. Although, their integration is as illegal as their settling in the town, these refugees found some kind of home in Kassala. However, it is difficult to say if this home is temporary or permanent. The issue of economy, highlighted by Kibreab, is a significant factor for people not chose to return migration. What is interesting is the fact that despite their illegal status, the refugees in Kassala are adjusted and work together with the host community in both economic and social activities of the town.

What factor(s) facilitated the cooperation of the groups in this community? The similarity of their Islamic values, and their clan systems seem to have contributed in narrowing the gap between the refugees and their hosts. Moreover, through daily acquaintance with the local people the refugees in Kassala improved their knowledge of Arabic, the language of the Koran. Nonetheless, I would hesitate to accept the position of those who claim that their increased knowledge of Islam means that they have become fundamentalists. However, they do object to the women's compulsory military service in Eritrea after they have come to know the political system in Sudan where the practice is different. The refugees' own reason for choosing to stay in Sudan rather than return to Eritrea is simple – they do not identify with the present political and social setting in Eritrea. In their new home they are accepted by the locals, they participate in the life of the community. They do not feel equally welcomed by those who support the regime in Eritrea. Home for these people is where they can feel welcome and identify with the local people.

The twenty-four key informants in my study do not constitute a representative sample of all Eritreans in the Kassala region. Strictly speaking it is therefore difficult to generalize from this data set. However, since I met a great many more people in casual conversations in the market, in people's homes, in coffee-shops and various public spaces the general impression that I got is that most of the Eritrean refugees who have remained in Sudan are predominately Moslem (the dominant religion in the study area), they are from the lowlands of Eritrea and many are members or former members of the ELF. This indicates that there is a link between refugees who sympathize with the ELF, refugees from the lowlands, refugees who share values and norms with the local people and refugees who are Moslem and opponents to the present regime in Eritrea.

Notes:

1. Family man who left Eritrea in 1960. He had then finished junior high school. He belongs to the Tigre clan and is an active member of the ELF. He works as guard and resides in Abukamsa. The interview took place in Kassala market and was held in Tigre and thereafter translated into English by me, Kassala, 2002.

2. Housewife who left Eritrea in 1980 after having finished junior highs school. She is a Christian and belongs to the Tigrinya clan. She resides in Abukamsa. The interview took

place her residence and was held in Tigrinya mixed with Arabic. It was translated into English by me, Kassala, 2002.

3. Family women who left Eritrea in 1977, She is head of the household, illiterate and lives on remittances. She is a Moslem and belongs to the Blin clan. She resides in Muraba'at. The interview took place in her house and was held in Tigre. It was translated into English by me, Kassala, 2002.

4.. Female, born in Eritrea but raised in Sudan. She is single and a university graduate. She is a Moslem, and belongs to the Blin clan. She lives with her parents in Muraba'at. The interview took place in Muraba'at in the UNHCR RSD registration centre and was conducted in Arabic. It was translated by me into English, Kassala, 2002.

5. Female, housewife.She has finished junior high school and lives partly on remittances. She belongs to the Blin clan. She resides in Abukamsa,. The interview took place in the UNHCR registration centre and was held in Arabic. It was translated by me into English, Kassala, 2002.

6. Female, housewife who has finished junior high school. She is a Moslem who belongs to the Tigre clan. The interview took place in Muraba'at in her home. It was held in Tigre and translated into English by me, Kassala, 2002.

8

SUMMARY AND CONCLUSION

This chapter is summing up the factors that influ ence the decision making regarding return migra tion.

Why Return Migration is not an Accepted Solution

The aim and focus of this study has been to examine the decision-making process of Eritrean refugees in Kassala town regarding return migration.

One of the factors that this study aimed to examine was the political question regarding return migration. This is also the contribution of this study to migration studies. For instance as indicated in chapter five it is clear that recognition and political inclusion are important factors when refugees consider return migration. Having said that, this study advocates the inclusion of the political factor in the policies of return migration. Moreover, this study points out that what may appear to be political in one instance may another time be social and vice versa. Beside this, this thesis shows that the refugees' failure to return is linked to political conditions in the country of origin as well as to the socioeconomic conditions in Sudan and the hope of resettlement in countries of the north and Australia.

Return migration is perceived by most of the official institutions involved with refugees as the most durable solution to the global refugee problem. If the policy of UNHCR is to succeed, the view of the refugees to the policy must be an important consideration. A method is thus required that measures the reactions expected of the refugees to the institutional implementations. The return migration decision-making process is difficult. The recognition of this difficult process has been an important factor in the choice of the data collecting method adopted in this thesis. The method adopted in this study has been to examine how feasible it is to implement the return policy in view of the circumstances of the Eritrean refugees in Sudan and their own understanding of what is best for them. As pointed out in part one, being an Eritrean refugee myself and engaging in the daily lives of people belonging to this community, observing their interactions and activities, has not been problem free. Although, cultural capital can be a positive thing, it can also be a barrier to a clearer appreciation of the concerns of international organisations, states and outside researchers. The difference between the insider researcher and the outsider researcher is based on the point of departure of each. While outsiders frame their questions from the point of view of either the UN or even that of the Sudanese government, the insider's questions depart from how the refugees perceive their situation. What makes this research different from previous studies is that the questions are conceived of from the point view of the refugees.

The study shows that the situation of these refugees is more complex than return migration policy might imply. For instance, in May 2002 the UNHCR announced the termination of refugee's status to all Eritrean refugees worldwide. This took place while Eritreans were arriving in Sudan on a daily basis. It appears that the agency did not reconsider its position even though the Eritrean refugees continued to stream in after the announcement as they had done before it. The UNHCR's reaction to the new situation made it appear as an inflexible agency out to pursue a policy rather than to evaluate the matters of this wave of refugees.

One of the complexities that the return policy does not seem to acknowledge is the changes and experiences that the refugees have undergone after many years in exile, in quite a few individual cases amounting to several decades.

This study indicate despite commonly held views about refugee assistance, the individual's decision whether to return or not is composed of multiple and complex political, economic and social issues that continuously change and shift. At least in the case of the Eritrean refugees in Kassala town, the question of their eventual return to their country of origin after the elimination of the factors that prompted them to flee, has to do with the political changes that have taken place in Eritrea, but also with intersecting factors linked to the country of asylum and personal changes refugees undergone in exile.

Are the Refugees Likely to Return to Eritrea?

The honest answer is that nobody knows. If one asks the refugees, including those who have found some kind of homes, they say that they will definitely do so. When probed as to when they intend to return, they say 'when there is democracy.' The question, however, is whether these answers are to be taken at their face values? Is it difficult to assume that all refugees in Kassala would return to Eritrea once the changes that they are asking for happen? Being rational human beings, the refugees will calculate very carefully the available possibilities in the country of exile as well as in the country of origin. Moreover, staying in Sudan also enables those with political ambitions to overthrow the Eritrean government, which they oppose strongly. .

The original factors that prompted Eritrean refugees to flee were eliminated many years ago, and yet a considerable share of those who fled due to the war of independence have remained in Sudan. The principal reasons coming out of this study that account for the Eritrean refugees' decision not to return to Eritrea are:

- √ Exclusion of the ELF from power sharing; and an unfavourable political situation in the country of origin
- √ Human rights violations committed by the present Eritrean regime
- √ Unwillingness to participate in the open-ended national service
- √ Membership in one of the exiled political organisations opposed to the Eritrean government
- √ An extended stay in exile; for many running into several decades. GA:overlaps the following two
- √ Integration in the host country at the individual level
- √ Social and economic changes experienced in exile
- √ Possession of transnational networks and lack of proper refugee status in the country of exile encourage attempts to find resettlement in countries of the north and Australia

These self-settled Eritrean refugees in Kassala town from whom my respondents were selected, present themselves as Eritrean refugees even though they have lived for decades in the host country. However, their maintenance of Eritrean identity does not mean that they are ready to return on any conditions. This is because the reasons why they fled to Sudan and the reasons why they have remained there differ. They were predominantly former or present members of the ELF. Indication is that the majority of the remaining Eritrean refugees in Sudan are Moslems who are historically linked to the ELF. It is interesting that a UNHCR memo from 2002 shows that the large majority of the refugees who returned to Eritrea were Christians. The pattern of the decisions made about return migration might be construed in many ways. One reason why the Eritrean Moslems have remained in Sudan while the Christians have returned is possibly that the Eritrean Moslems in Sudan fear that the Moslem community will be at a social, economic and political disadvantage in Eritrea.

Some of the refugees in Kassala town are either members or sympathisers of the ELF, which is forbidden in the one party state of Eritrea. These refugees are in all likelihood voicing the general opinion of their fellow citizens when they say that they want a multiparty Eritrean state that respects democratic values. Until this happens, they are going to continue their political struggle in exile. Many of the respondents openly state that they strongly dislike the government in their country of origin. The reasons why they do so vary. Some disfavour the government because it ejected the ELF from the Eritrean political arena. Others disfavour the government because in their minds its actions appear to be anti-Islamic. Prior to their displacement, many of the respondents lived in the parts of Eritrea where the ELF operated. These refugees sympathise with the ELF in exile while some have actively continued their membership of the ELF. These informants view the faction as their political representative that safeguards their interests. Therefore, for them to return to Eritrea while the faction is not included in the Eritrean political area is impossible. Even Eritreans in Kassala who did not belong to the group of active ELF supporters still held the opinion that the ELF has a right to share power. Many of them were angered by its exclusion from power, and this has a bearing on their decision not to return to Eritrea. Moreover, the study shows that ELF is not only a political faction; it also represents something of a substitute for family available in the country of exile for some of the respondents in this study. This indicates that motives for staying that once were political can at some other time and in a different context transform into social motives, and vice versa. Social and economic issues have dominated explanations given by earlier studies on return migration. Looking at the results of this study and comparing them to those of Kibreab (1996), the difference is mainly that Kibreab does not stress the political factor as vital to return migration while I have found that it does play an important role. Although social and economic factors are important for the decision making of the people covered by this study, political considerations are equally important. The role that politics actually play for people's acceptance of return migration as a solution to their situation is one of the findings of the study.

All Eritreans, men and women alike, are required to participate in the open-ended national service. This requirement also discourages many émigrés from returning to Eritrea. Many young Eritreans arrive in Sudan after having fled from the national service. They face the risk of serious punishment should they return. A large majority of the refugees in Kassala appear to be strongly opposed to the participation of women in the national service. Many rumours are spread about the sexual abuse that women who participate in the national service are subjected to. There are stories of young women who return to their parents pregnant. Whether there is some kernel of truth or not in these rumours, I cannot tell. The important social fact is that if people believe them to be true, this will serve as a strong motive not to return.

This study indicates that once people flee and settle elsewhere for a lengthy period of time they undergo changes in terms of identities, life styles, frames of reference and social position. The whole sociocultural structure of the receiving society stages the conditions of everyday life to which newcomers are exposed and will have to relate and adapt to.(Whether refugees accept their migrant situation, either by acquiring citizenship or by regularising their stay by other means on the one hand, or whether they return after the elimination of the factors that forced them to flee on the other hand, depends upon the comparative advantages of the countries of asylum and that of origin.) Some of my respondents have undergone profound social and economic changes in exile. Prior to their displacement most of them used to live in rural areas in Eritrea for the simple reason that urban development in the Moslem dominated regions had not come very far when the war of independence started. In Sudan, however, they have grown accustomed to urban life conditions. Should they return they want to settle in urban areas and lead the same kind of life as they have led in Sudan?

Livelihood issues are not limited to income earning or acquisition of property alone. For some women, the refugee experience entails release from restrictions imposed by patriarchal practices. For the respondents of this study this change was not viewed as a good thing. Instead, it was regarded as a burden and an additional responsibility, which contradicts the commonly held view by many migration researchers that migration frees female refugees from the traditional roles and control, and stimulates them to challenge discriminatory values and social norms. One of the consequences of the breakdown of the old structures and old systems of livelihood is that the distinction between the feminine and masculine domain becomes either eroded or blurred. The loss of men's breadwinning ability enables many women to assume the role of breadwinners, which is a change in gender roles.

A further aspect that appeared in this study is that, although refugees upon their arrival in the host country are in a precarious situation this study shows that the Eritrean refugees soon gathered their strengths and creative skills and managed to survive on their own, thus contradicting the myth that refugee are weak and helpless. Based on experiences of the refugees in Kassala one should perhaps be careful not to stretch their temporary setback into a permanent state of weakness, helplessness and dependency. These refugees have undergone many changes in terms of working life, social life, gender roles, political achievements and frustrations, and family relations. In the meantime, their country of origin also underwent dramatic changes from being a country under foreign occupation to becoming an independent state. The demand for independence was the cause of the thirty-year armed struggle between Eritrean freedom fighters and the Ethiopian army. Most Eritreans view the liberation of Eritrea as a positive step. In spite of the achieved national independence, many refugees in Sudan are reluctant to return to Eritrea. It is clear that while they are positively disposed towards an independent state, it does not seem to offer sufficient motivation for them to return. This indicates that a change of regime, or achieved independence are not the only factors that can get the Eritrean refugees in Kassala to return to Eritrea.

A considerable proportion of the respondents in Kassala town are linked by transnational networks to Eritrean Diaspora. Most of the respondents have relatives in Europe, North America and Australia. Not only do these refugees receive remittances but they also want to join their families who are settled in these countries. Although resettlement to countries of the north is associated with difficulties, this is something that practically everyone hopes for. Some of my respondents express greater determination than others to be resettled against all odds.

My study shows that migrating to a third country is one of the factors that affect the decision making of the informants. However, in the process of analysing the information the respondents have given, new questions have arisen. One important question concerns the many implications of third country resettlement.

What has become clear from the point of view of the Eritrean refugees in Sudan is that the third country option means resettlement in countries of the north and Australia, and not in an African third country. It seems that the motivating drive for their hope of resettlement is the belief that to be settled in one these countries affords them the right to be integrated, and even to become citizens of those countries. Moreover, this study shows that those refugees in Kassala, who have relatives in the countries of the north, receive money remittances from their kin. These remittances are vital for the survival of the Kassala refugees. The respondents express a desire to re-establish or maintain their social networks with relatives farther a field. In the relationship of the refugee network, the financial remittances depend on the socio-economic standing of the sender in the north. An important question that could be explored in future research is the civic and economic standing of the Eritreans in the north who send the remittances. I would like to pursue this question with respect to the Eritrean refugees in Sweden and Australia. How economically well off are these refugees compared to their kin in the remittance receiving countries? What civic status do they have compared to their relatives in Kassala? What are the legal and experienced differences between the two groups of refugees? What are the implications for local integration and for citizenship?

What Messages does my Study Send to Communities in Eritrea and in other parts of Africa?

As seen in the previous chapters, analysing refugee movements both as mobility and as organisation requires the consideration of historical, economic, social and political factors. Although, all these aspects are interrelated to each other, the main points are that without democracy, rule of law as well as economic and social development, the policy of return migration will be very difficult to implement. Migration, forced or voluntary, touches upon central definitions of governance.

Diplomatic solutions among the concerned countries are the key to return migration. That is what the international community and those who care for refugees have to concentrate on to solve the refugee problem.

Many of the political problems haunting African

countries can be put down to nation formation processes. We have to distinguish between nation, country and state (Westin, 1999). The nation represents the people with their culture, language, religion and history. The country represents the land with its resources, economy and geography. The state represents governance in terms of legislation, rule, administration and political control (ibid). It is when these three aspects do not function together that conflict and disagreement between citizens and those who are in power takes place. It is also due to this dissatisfaction that people decide to leave their area of origin. What messages does my study send to communities in Eritrea and in other parts of Africa?

Human rights violations conducted by dictatorial regimes in the region of the horn of Africa are the reason why many people leave their countries of origin. This is why today this region is one of the areas that generate and receive most refugees on the continent. The countries of this area are known for their internal conflicts and many people from these areas were/are displaced. The refugees that sought asylum in these countries live in a state of limbo. They cannot return to their countries of origin or migrate further. The governments in this region oppose and fight each other. As observed recently between Somalia and Ethiopia they interfere in each other's internal affairs. Despite political tension at the governmental level the people of these countries have shown a remarkable hospitality, tolerance and kindness towards the refugees. This indicates that the conflicts do not affect the local people's attitude within the region. We have a situation of Eritrean refugees in Sudan, Somali refugees in Sudan, Sudanese refugees in Eritrea and Eritrean refugees in Ethiopia. What does this tell us? Does this indicate that the regimes in this region do not know what these conflicts mean to their citizens? As this study shows Eritrean and Sudanese citizens of the border regions are connected through common religion, ethnicity, economic interests and culture. The border communities want to be able to move between Sudan and Eritrea and many persons I spoke to want to live in both countries. Traditional "cross-border" contacts that have taken place for centuries for purposes of trade and exchange are now regarded as unrealistic unless special arrangements between the two countries are settled.

Such aspects challenge the notion of nation, state and territoriality. As many studies show, the root cause of the conflicts in Sub-Saharan Africa dates back to the boundaries that were settled by the colonial powers at the conference of Berlin in 1878. The boundaries were drawn without any social consideration to the local people. Thus, the borders of most of these countries cut right through territories of distinct ethnic groups. These borders therefore lack social logic, a fact that can be observed in the way refugees are treated by their host population.

The state is the main actor that controls territory also confers citizenship rights. The state imposes obligations to its citizens, namely to protect the country in case it is threatened by an outside enemy. As in the case of Eritrea, the feeling of threat can also be connected with internal affairs that arise due to bad governance or human rights violations. Although, violations against its people are a common problem in the region of the Horn of Africa, each country does it different ways. Today almost all the countries disregard the universal human rights conventions.

Final Comments

The present Eritrean state regardless of its appearance oppresses whoever opposes it and embraces whoever serves its interests. However, the general perception among the respondents of this study was that the government in Eritrea serves the interest of Christians at the expense of Moslems; the government of Eritrea needs to reassure fellow Eritreans in exile that their perceptions are unfounded. Otherwise, Eritrea will continue to produce refugees and those who are in exile will be reluctant to return.

This is so despite the fact that during the Eritrean armed struggle both the ELF and the EPLF had a policy to prevent sectarianism based on religion, region, clan and ethnicity. The reason why the two fronts suppressed these sentiments and inclinations was to promote national unity and Eritrean national identity. One of the key factors that contributed to the success of the independence war was because Eritreans were able to relegate their differences and particular interests to the background. The Eritrean community has to work on the socio-political differences that have caused so much alienation and dispute. However, in the post-independence period, Eritrean people do not seem to be held back from expressing sectarian views.

A solution to the Eritrean refugee problem cannot be reached without resolving the conflicts between the governments of the Horn of Africa, especially Eritrea and Sudan. Sudan was one of the countries that provided asylum to hundreds of thousands of Eritrean refugees. The Eritrean refugees are also appreciative of the hospitality that Sudan has offered. The livelihoods of the border communities in both countries are dependent on cross-border economic activities and therefore it is important that the borders remain open to allow the movement of goods and services between the two countries. This will contribute substantially to the solution of the Eritrean refugee problem. Cross-border movement of people will not cease once the refugees return home. People will continue to move back and forth in search of employment, water, grazing land and different forms of income-generating activities, including trade. Open borders are what the refugees want and it is important that the two governments consider this in their diplomatic relations.

REFERENCES

Alinia, M (2004), *Spaces of Diasporas. Kurdish identities, experiences of otherness and politics of belonging.* Göteborg studies in sociology no, 22.Göteborg: Department of Sociology, Gothenburg University.

Allen, T and Morsak, H (1994), *When refugees go home.* Geneva: UNRISD.

Allen T and Turton, D (1996), 'Introduction', in Allen, T. (Ed.) *In search of cool ground.* p. 1-22 Geneva: UNRISD.

Ammar, M (1992), *Eritrean root causes of war and refugees.* Baghdad: Wleed Mahdi Sindbad Printing Press,

Bascom, J (1998), *Losing place, refugee population and rural transformation in east Africa.* New York: Berghahn Books.

Bayoumi, A (1999), *The socio-economic survey of Eritrean and Ethiopian Refugees in Eastern Sudan.* Khartoum: MRC.

Berry, J.W (1990), 'Psychology of acculturation'. In J.J. Berman (Ed.), *Nebraska symposium on motivation 1989*: Cross-cultural perspectives, p. 201–234. Lincoln: University of Nebraska Press.

Berry, J. W (1997), 'Immigration, Acculturation and Adaptation'. *Applied Psychology: An International Review* 46, (1): 5-68.

Black, R and Koser, K (Eds.) (1999), *The end of the refugee cycle, refugee repatriation and reconstruction*. Refugee and forced migration studies. Volume, 4. New York: Berghahn Books.

Bondestam, L (1989), *Eritrea med rätt till självbestämmande*. Göteborg: Clavis.

Brekke, J (2004), *While We are Waiting. Uncertainties and Empowerment among Asylum Seekers in Sweden*. Oslo: Institute for Social Research.

Bulcha, M (1988), *Flight and integration. Causes of mass exodus from Ethiopia and problems of integration in the Sudan.* Ph.D. thesis. Uppsala: Nordic Africa Institute.

Castles, S and Miller, M.M (1998), *The Age of Migration*. London: Macmillan Press.

Connell, D (1998), 'Strategies for change: Women and Politics in Eritrea and South Africa'. *Review of African political economy,* 76, 189-206.

Dias, J.A (1993), *Choosing integration: a theoretical and empirical study of the immigrant integration in Sweden*. Uppsala: Department of Sociology, Uppsala University, Ph.D. thesis.

Dolyal, L (1996), *What makes women sick? Gender and the political economy of health*. London: Macmillan Press.

Duncan, S (1994), 'Theorising differences in patriarchy'. *Environmental and planning,* 26, 1177-1194. London: Gender institute, London, school of Economics.

Durkheim, É (1984), *The Division of Labour in Society*. London: Macmillan.

Eastmond, M and Öjendal, J. (1999), 'Revisiting a "Repatriation Success": The case of Cambodia', in Black, R. and Koser, K. (Eds.), *The End of the Refugee Cycle.* Refugee and forced migration studies. Volume, 4. P, 38-55, New York: Berghahn Books.

Edward, K (2000), 'South Sudanese refugee women: questioning the past, imagining the future'. In *Women's Rights and Human Rights international perspective*, Grimshaw, P, Holes, K. (Eds.), p. 272-288. Gordonsville, VA.: Palgrave Macmillan.

Eyle, J. and Smith, D (1988), *Qualitative methods in human geography*. Cambrige: Polity Press.

Favali, L. and Pateman, R (2003), *Blood, land and sex. Legal and political pluralism in Eritrea*. Bloomington: Indiana University Press.

Fellesson, M (2003), *Prolonged exile in relative isolation, long-term consequence of contrasting refugee polices in Tanzania.* Uppsala: Department of Sociology, Uppsala University.

Ghanem, T (2003), *When forced migrants return "home". The psychosocial difficulties returnees encounter in the reintegration process.* Oxford: Queen Elisabeth House International Development Centre, University of Oxford.

Gilbert, A & Gugler, J (1982), *Cities, Poverty and Development, Urbanisation in the third world*. London: Oxford University Press.

Gordon, M (1964), *Assimilation in American Life*. New York: Oxford University Press.

Goutom, E (1980), *Adoption and integration.The case of the Eritrean refugees in the three towns, 1974-1979*. Khartoum: Department of Geography, University of Khartoum.

Göte, H (2003), *Building a New State. Lessons from Eritrea*. Oxford: Oxford University Press.

Habete Selasssie, B (1989), *Eritrea and the United Nations and other essays*. Trenton, NJ: Red Sea Press.

Hammond, L (1999) 'Examining the discourse of repatriation: towards a more proactive theory of return migration', in Black, R. and Koser, K. (Eds.), *The end of refugee cycle.* Refugee and forced migration studies. Volume, 4. p, 227-244. New York: Berghahn Books.

Hannerz, U (1996), *Transnational Connections*. London: Routledge.

Hassanen, S (1997), *Female Genital Mutilation in East Africa with special reference to the situation in Sudan.* Thesis for a Masters Degree in Public Health. Umeå: Department of Epidemiology and Public Health, Umeå University.

Hassanen, S (2000), The effect of migration among the Blin people in Melbourne from a gender perspective. Unpublished paper. Umeå: Department of Geography, Umeå University.

Hassanen, S (2002), Gender and migration. Some reflections about the situation of migrant women in host countries. Unpublished working paper. Stockholm: Department of Human Geography, Stockholm University.

Hendrie, B (1996), 'Assisting Refugees in the Context of Warfare', in Allen, T. (Ed.) *In search of cool ground*. pp. 1-22, 35-44 Geneva: UNRISD.

Hondagenu, S (1994), *Gender transitions. Mexican experiences of immigration.* Berkeley: University of California Press.

Horst, C (2003), *Transnational Nomads. How Somalis cope with refugee life in the Dadaab camp of Kenya*.Ph.D thesis. Amsterdam: Department of Anthropology, Amsterdam University.

Ibrahim, O (2003), *Proportion of the Voluntary Repatriation Rejection of the Eritrean Refugees in Kassala state Camps*. Kassala: Alneelain University.

Indira, D (1999), 'Not a "Room of one's Own": Engendering Forced migration Knowledge and Practice'. in Indira, D. (Ed.), *Engendering Forced migration Theory and Practice.* Refugee and forced migration studies. Volume, 6. p. 1- 22, New York: Berghahn Books.

Jupp, J (Ed.) (2001), *The Australian People. An Encyclopaedia of the Nation, Its People and their Origins.* 2nd Edition. Cambridge: Cambridge University Press.

Karadawi, A (1978), *Urban refugees in the Sudan 1975-1978*. Khartoum: (COR) Office of the commissioner for refugees, Ministry of interior.

Karadawi, A (1999), *Refugee policy in Sudan, 1967-1984*. New York: Berghahn Books.

Kibreab, G (1983), *Replications on the African refugee problems: A critical analysis of some basic assumptions.* Research report, No, 67. Uppsala: The Scandinavian Institute of African Studies.

Kibreab, G (1987), *Refugees and development in Africa. The case of Eritrea.* Trenton NJ: Red Sea Press.

Kibreab G (1989), *Refugee settlement and development in Africa. A study of organised settlement for Eritrean refugees in eastern Sudan* 1996-1983. Uppsala: Department of economic history, Uppsala University.

Kibreab G (1990), *The refugee problem in Africa*. Asmara: Research information centre on Eritrea (RICE).

Kibreab, G (1996a), *People on the edge in the horn. Displacement, land use and the environment in the Gedarf region, Sudan.* Cincinatti OH: Ohio University Press.

Kibreab, G (1996b), *Ready and willing but still waiting, Eritrean refugees in Sudan and the dilemmas of return*. Uppsala: Life and peace institute.

Kibreab, G (1996c) 'Eritrean and Ethiopian refugees in Khartoum. What the eye refugees to see', *African review,* 39, 31-178.

Kibreab, G (2003), *Displaced communities and the reconstruction of livelihood in Eritrea.* Oxford: Oxford University Press.

King, R (1997), *Mass migration in Europe, the legacy and the future*. London: Belhaven Press.

Kuhlman, T (1994), Asylum or aid? The economic integration of Ethiopian and Eritrean Refugees in the Sudan. *African Studies centre research series (*TY00222719, pub (Aldershot): Avebury.

Kuhlman, T (1990), *Burden or boom?* A study of Eritrean refugees in the Sudan. Amsterdam: Anthropological studies, University of Amsterdam.

Kunz, E (1981), 'Exile and resettlement, refugee theory'. *International Migration Review,* 15, 42-51.

Kvale, S (1997), *Interviews. An introduction to qualitative research, interviewing.* Thousand Oaks, CA.: Sage.

Malkki, L (1995), *Purity and exile, violence, memory and national cosmology among the Hutu refugees in Tanzania*. Chicago: University of Chicago Press.

Malmberg, G (1997), 'Time and Space in International Migration', in Hammar, T., Brochmann, G., Tamas, H. and Faist, T (Eds.) *International Migration Immobility and Development. Multidisplinary Perspective,* p. 1-19 Oxford: Berg.

Mcspadden, L (1999), 'Contradictions and control in repatriation, negotiation for the return of 500.000 Eritrean refugees', in Black, R. and Koser, K. (Eds.), *The end of refugee cycle.* Refugee and forced migration studies. Volume, 4. p. 69-84. New York: Berghahn Books.

Mcspadden, L (2000), *Negotiating return: Conflict and Control in the repatriation of Eritrean refugees*. Uppsala: Life and Peace Institute.

Mohammad, R (2001), '"Insiders" and/or "outsiders:" Positionality, theory and praxis", in Limb, M. and Dwyer, C. (Eds.), *Qualitative Methodologies for Geographers. Issues and debates*. p. 101-117. Oxford: Oxford University Press.

Moussa, H (1993), *Storm and Sanctuary. The Journey of Ethiopian and Eritrean Women Refugees.* Dundas Ontario: Artemis Enterprises.

Moussa, H (1995), 'Caught between two worlds. Eritrean women refugees and voluntary repatriation'. In Sorensen, J. (Ed.) *Disaster and Development in the Horn of Africa.* London: St Martin's Press.

Nachmias, F. and Nachmias, D (1996), *Research Methods in Social Sciences*. London: St. Martin's Press.

Naty, A (2002), *Potential conflicts in the former Gash-Setit region, Western Eritrea threats to peace and security.* Asmara: Department of Anthropology and Archeology, University of Asmara.

Nieeswiadadomy, R (1998), *Foundation of nursing research,* third edition. Texas Women's University Collage of Nursing. Stamford, CONN.: Appleton & Lange.

Pateman, R. (2001), 'Eritreans'. In Jupp, J (Ed.), *The Australian People. An Encyclopedia of the Nation, Its People and Their Origins*, p. 344-345. Cambridge: Cambridge University Press.

Peil, M (1998), Consensus, Conflict and Change: A Sociological Introduction to African societies. Nairobi: East African Educational Publishers.

Pool, D (2001), *From Guerrillas to Government. The Eritrean People's Liberation Front*. Eastern African studies. Oxford: James Carrey.

Ramazanoglu, C. and Holland, J (2002), *Feminist Methodology. Challenges and Choices*. London: Thousand Oaks.

Rogge, J (1994), 'Repatriation of refugees', in Allen, T. and Morsink, H. (Eds.) *When refugees go home*. p. 14-49. Geneva: UNRISD.

Rogge, J and Akhol, J (1989), 'Repatriation: Its role in resolving Africa's Refugee Dilemma'. *International migration review*, p. 184-200. New York: Centre for migration studies.

Silberschmidt, M (1999), W*omen forget that men are the masters. Gender antagonism and socio-economic change in Kisi district in Kenya*. Uppsala: Nordic Africa Institute.

Sorensen, N., Van Hear, N. and Pedersen, P (2002), The migration development nexus evidence and policy options. Working paper 02.6 Copenhagen: Centre of development research.

Storti, C (2001), *The art of coming home*. London: Intercultural Press.

Walby, S (1990), *Theoretizing Patriarchy*. Padstow Cornwall: T. J. Press Ltd.

Westin, C (1996), 'Migration patterns', in Haour-Knipe, M. and Rector, R. (Eds.) *Crossing Borders. Migration, Ethnicity and Aids*. p. 15-30. London: Taylor and Francis.

Westin, C (1999), 'Regional analysis of refugees movement. Origin and response'. In Ager, A (Ed.) *Refugees. Perspectives on the Experiences of Forced Migration*. p. 24-45. London: Cassell.

Westin, C (2005), 'Diversity, national identity and social cohesion'. AMID Working Paper Series 41/2005. Aalborg: AMID.

Wijbrandi, S (1986), *Organised and spontaneous settlements in Eastern Sudan. Two case studies on integration for rural refugees.* Amsterdam: Free University of Amsterdam.

Zolberg, A.R., Suhrke, A. and Aguayo, S (1989). *Escape*

from Violence. Conflict and Refugee Crisis in the Developing World. Oxford: Oxford University Press.

References from Journals

Al-Rashedd, M (1994), 'The myth of return: Iraqi Arab and Assyrian refugees in London'. *Journal of Refugee Studies,* 7, 199-219.

Bascom, J (1989), Social differentiation among Eritrean refugees in eastern Sudan: The case of Wad el Hileau', *Journal of Refugee Studies*, 2, P, 419-440.

Bascom, J (2005), 'The long "last step". Reintegration of repatriates in Eritrea'. *Journal of Refugee Studies,* 18, 165-181.

Graham, M. and Khosravi, S (1997), 'Home is where you make it. Repatriation and Diaspora among Iranians in Sweden'. *Journal of Refugee Studies,* 10, 115-132.

Habib, N. (1996), 'The search of home'', 1996, *Journal of Refugee Studies,* 9, 96-102.

Kibreab G (1999), 'Revisiting the debase on people and place'. *Journal of Refugee Studies*, 12, 384-410.

Kibreab, G (2002), 'When refugees come home. The relationship between stayees and returnees in post conflict Eritrea'. *Journal of Contemporary African Studies*, 20, 384-410.

Kok, W (1989), 'Self settled refugees and the socio-economic impact of their presence on Kassala, Eastern Sudan'. *Journal of Refugee Studies,* 2, 419-440.

Koser, K (1996), 'Changing agendas in the study of forced migration: A report on the fifth international research and advisory panel meeting'. *Journal of Refugee Studies,* 9, P, 353-366.

Koser, K (1997), 'Information and repatriation the case

of Mozambican refugees in Malawi'. *Journal of Refugee Studies,*10, 1-16.

Kuhlman, T (1991), 'The economic integration of refugees in developing countries, a research model', *Journal of refugee studies*, 4, 1-20.

Sorensen, J (1990), 'Opposition, exile and identity. The Eritrean case'. *Journal of Refugee Studies*, 4, 298-318.

Stepputat, F (1994), 'Repatriation and the politics of space: the case of the Maya Diaspora and return movement'. *Journal of Refugee Studies,* 7, 175-185.

Warner, D (1994), 'Voluntary repatriation and meaning of return home. A critic of liberal Mathematics'. *Journal of Refugee Studies,* 4, 160-174.

References from UNHCR and other Home Pages

Al-Sharmani, M (2004), Refugee livelihoods. Livelihood and diasporic identity constrictions of Somali refugees in Cairo. www.UNHCR.ch date of access, April 2, 2006.

Bakewell, O (1999), Returning refugees migrating villagers? Voluntary repatriation Programmes in Africa reconsidered, www.UNHCR.ch date of access, October 15, 2005.

Chimini, B (1999), From resettlement to involuntary repatriation, towards a critical history of durable solutions to refugee problem. www.UNHCR.ch, Date of access, June, 12, 2002.

Crisp, J. (1994), The local integration and local settlement of refugees, conceptual and historical analysis. www.UNHCR.ch date of access September, 14, 1999.

Crisp, J. (1999), A state of security: the political economy of violence in refugee-populated areas of Kenya. www.UNHCR.ch date of access August 5, 2001.

Crisp, J. (2000), Africa's refugees patterns, problems and policy challenges. www.UNHCR.ch date of access 3 May, 2003.

Crisp, J. (2001) Mind the gap! UNHCR humanitarian assistance and the development process. www.UNHCR.ch date of access 12, March, 2003.

Crisp. J. (2002) Protracted refugee situation, some frequently asked questions. www.UNHCR.ch date of access February 2, 2006.

Crisp, J. (2003), No solution in sight: the problem of protracted refugee situation in Africa. www.UNHCR.ch date of access, April 3, 2003.

Crisp, J. (2004), The local integration and local settlement of refugees: Conceptual and historical analysis. www.UNHCR.ch date of access, April, 2003

Horst, C (2003a), Vital links in social security. Somali refugees in the Dadaab camp in Kenya. www.UNHCR.ch date of access, October, 12 2005.

Jacobsen, C. & Landau, L (2003), Researching refugees some methodological and ethnical consideration in social science and forced migration. www.UNHCR.ch date of access, 10, December 2004

Jacobsen, K. (2001), The forgotten solution, social integration for refugees in developing countries. www.UNHCR.ch date of access, November,6 2004.

Marvis, L (2002), Human smugglers and social networks: Transit migration through states of former Yugoslavia www.UNHCR.ch date of access, November 20, 2006.

Morrison, J. and Crossland, B (2001), The trafficking and smuggling of refugees. The end of game in European asylum policy. www.UNHCR.ch date of access, December 5, 2004.

Riak, A (2004), Remittances as unforeseen burden, considering displacement, family in addition, resettlement context in refugee livelihood and well-being, is there anything states. Or organisation can do. www.gcim.org date of access, March 7, 2006.

Sorensen, N (2005), Migrant remittances, development and gender. Copenhagen: Dansk institute for international studier. www.diis.dk date of access, March 25, 2006.

Van Hear, N (2003), From durable solutions to transnational relations: home and exile among refugee diasporas, www.UNHCR.ch date of access February 12, 2005.

Amnesty International and Human Rights Reports

"You have the right to ask. Government resistance scrutiny on human rights." Amnesty International report 2003. http://www.amnesty.org Date of access, March, 8, 2006

Eritrea: Refugees Involuntary repatriated from Libya. Human rights watch report 2004. http://www.hrw.org Date of access, February 4, 2005

QUESTION GUIDE

Background information

Date and motive of migration. Application for refugee status.
Age, Sex, Religion, Clan affiliation, Political affiliation, marital status, region of origin in Eritrea, Educational background, Family size, Neighbourhood of residence in Kassala, Knowledge in Arabic,

Livelihood issues and contacts with the local people

How do the refugees accomplish their financial needs? How is the employment situation in Kassala, what kind of jobs are available for refugees, how do refugees manage to be employed?

The role of having contact with the local people regarding livelihood.

How do the refugees feel about the local people and what kind of relation do they have with them? Do they have similar contacts with other Eritrean refugees? How do they do these contacts and how much can the refugees depend on their relation with the local people? To the refugees who own their business and property in Kassala, how do they manage to do that and who helped them with the official papers? Those who receive remittances, who send it to them, how often they get it and in which country are those resided who sent to them.

About return to Eritrea and future plans

Why have the refugees remained until know in Sudan? What made them to decided to stay in Sudan, what do they need to decide to return to Eritrea or not, how do they feel about the local people's attitude towards them, What are the things that they learned from settling in Sudan and what are their future plans.

Discussion and Interviews with Refugee Groups in different Districts in Kassala and Refugees, COR and UNHCR Staff in the Town of Shewak and in Khartoum

√ End of August, 2002, interviews and discussion with personnel at the head office of COR in Khartoum, the conversation was held in Arabic.

√ Beginning of September 2002, Interview and discussion with one of COR staff who was responsible for the repatriation program in Khartoum. The conversation was held in Arabic in the Office of COR

√ September 2002, informal discussion with six Eritrea refugee women in Kassala. The discussion was held in Arabic in the house of one of them, in Abu-Khamsa district and the women belong to the Saho, Tigre, Jeberti and Blin clans. Out of the six, three were high school graduates. Two of them returned to Eritrea and then they returned back to Kassala, in 2000. The rest were resided in Kassala for more than 20 years.

- √ September 2002, discussion with representatives of the Bara clan in the registration centres for repatriation called Al-Sebil.
- √ In September 2002, Participation in repatriation campaigns in Kassala, and discussion with the refugees after meeting, the discussion was held in Tigre and the district is called Muraba'at.
- √ At the end of September, discussion with refugees in Kassala market. The discussion was held in Tigre.
- √ Visit to Girba prisoner in the beginning of October in order to meet refugees who were arrested on border area between Sudan and Ethiopia called Garagafe. To meet these people I went to the town of Girba
- √ In October, 2002 interview and discussion was held with the head of COR Kassala, the conversation was held in Arabic in the office of COR
- √ In October 2002, discussion with a group of refugee committee whom I met by chance in the COR office in Kassala. The discussion was held in Arabic and the area is called Al-kara.
- √ October 2002, Discussion with refugee women in the Girba refugee camp
- √ June 2003, interview and discussions with the head of COR office in the town of Shewak. The interview was held in Arabic
- √ June 2003, interview and discussions with the vice director of COR in the town of Shewak. The interview was held in Arabic
- √ June 2003, interview and discussions with one of UNHCR staff in Shewak, the Interview was held in English.
- √ June 2003, discussion with the refugee Committee in Wedsherifey refugee camp. The discussion was held in Tigre in the office of COR.
- √ June 2003, interview and discussion with the head of the refugee camp of Wedsherify, the conversation was held in Tigre in the office of COR

√ Mid of June 2003, a discussion with three differt groups of refugees of whom some work for UNHCR was held the in the town of Shewak. The discussion was held in Tigre Arabic and Blin head in the sub office centre of UNHCR.

Female and male respondents by gender, district in Kassala, clan affiliation, Political affiliation and date of interview

(Ka-Market=Kassala Market); (F=Female); (M=Male)

Case	Gender	District	Political Affiliation	Clan Affiliation	Religion	Date of interview
1.	F	Muraba'at	Ex ELF	Blin	Moslem	Oct,2002
2.	F	Ka-market	ELF	Tigre	Christian	Oct,2002
3	F	Abukamsa	ELF	Tigrinya	Moslem	Oct,2002
4	F	Muraba'at	ELF	Blin	Moslem	Oct,2002
5	F	Abuakamsa	ELF	Tigre	Moslem	Oct,2002
6	F	Ka-mark	ELF	Blin	Moslem	Oct,2002
7	F	Muraba'at	ELF	Tigre	Moslem	Oct,2002
8	F	Abuksamsa	-	Tigrinya	Christian	Oct,2002
9	F	Muraba'at	-	Tigre	Moslem	Oct,2002
10	F	Abuskamsa	Ex ELF	Tigre	Moslem	Oct,2002
11	F	Muraba'at	-	Tigrinya	Christian	Oct,2002
12	F	Muraba'at	-	Tigrinya	Christian	Oct,2002
13	F	Abuskamsa	-	Blin	Moslem	Oct,2002

Female and male respondents by gender, district in Kassala, clan affiliation, Political affiliation and date of interview

(Ka-Market=Kassala Market); (F=Female); (M=Male)

Case	Gender	District	Political Affiliation	Clan Affiliation	Religion	Date of interview
1.	M	Ka-Market	ELF	Blin	Moslem	Oct,2002
2.	M	Muraba'at	ELF	Blin	Christian	Oct,2002
3	M	Abukamsa	ELF	Tigre	Moslem	Oct,2002
4	M	Muraba'at	Ex ELF	Tigre	Moslem	Oct,2002
5	M	Abuakamsa	Ex ELF	Tigrinya	Christian	Oct,2002
6	M	Muraba'at	ELF	Tigre	Moslem	Oct,2002
7	M	Ka-Market	ELF	Tigre	Moslem	Oct,2002
8	M	Ka-Market	ELF	Tigre	Moslem	Oct,2002
9	M	Ka-Market	ELF	Tigre	Moslem	Oct,2002
10	M	Muraba'at	Ex ELF	Blin	Christian	Oct,2002
11	M	Abuskamsa Muraba'at	Ex EPLF	Tigrinya	Christian	Oct,2002

REPLICATED INFORMATION FROM THE UNITED NATIONS INTEGRATED REGIONAL INFORMATION NETWORKS (IRIN) AND UNHCR HOME PAGES

According to the sources these report do not necessarily reflect the views of the United Nations

Eritrea: Eritreans reapplying for refugee status in Sudan

[NAIROBI, 11 Feb 2003 (IRIN) - Tens of thousands of Eritreans are reapplying for refugee status in Sudan,

according to the UN refugee agency (UNHCR).

It said that more than a month after the 31 December deadline which ended refugee status for hundreds of thousands of Eritreans - most of them living in Sudan - dozens of legal teams are sifting through nearly 27,000 applications from Eritreans who want to remain in Sudan as refugees.

The applications - one per family - represent nearly 100,000 people living mainly in refugee camps and urban centres in Sudan.

"Some Eritreans say they cannot return home for fear of persecution because of their political affiliations or religious beliefs. Others say their marriage to non-Eritreans, particularly to Ethiopians, will place them and their families at risk if they return home," UNHCR said.

The legal teams recently completed screening applications in Gedaref and Wad Madani, and have shifted the emphasis to remote refugee camps in eastern Sudan before these are rendered inaccessible by the rainy season.

Meanwhile, some 32,000 Eritreans who have registered to go home are still waiting for the return operation to resume. Repatriation was suspended last June due to the rainy season and was expected to start again in October. But this was delayed by the closure of the Sudan-Eritrea border. UNHCR says it is trying to negotiate with both governments to allow the stalled operation to resume.

More than 100,000 Eritrean refugees have already returned home, some 50,000 of them with UNHCR assistance, the agency said. An estimated 223,000 Eritreans remain in Sudan, where many sought asylum more than 30 years ago. Some 92,000 of them are in refugee camps, while the rest are assumed to be in urban areas.

ERITREA: DEADLINE PASSES FOR ENDING REFUGEE STATUS

NAIROBI, 2 Jan 2003 (IRIN) - With the deadline for the cessation of refugee

status for Eritreans expiring on 31 December, the UN refugee agency (UNHCR) says thousands are seeking continued refugee status, while others have asked to be taken home or have applied to remain as immigrants.

Neighbouring Sudan hosts the largest number of refugees, and the number of Eritreans seeking interviews to determine their need for continued protection more than doubled in the two weeks before the deadline, UNHCR said.

"There has been a pick-up in registration for refugee status determination these last several weeks," said Ahmed Said Farah, UNHCR's representative in Sudan. Some 100,000 Eritreans have approached screening teams in the country.

"The mass information campaign to inform the Eritrean refugees of their options is active, and registration centres are open daily," Farah said.

UNHCR said Eritrean refugees were also asking to join repatriation convoys from Sudan, which are due to resume on Sunday after a six-month gap due to heavy rains and tension in Sudan's border region with Eritrea. More than 20,000 people have so far registered to return home once the convoys resume.

A total of 103,000 refugees, out of over 320,000 in Sudan, have returned since May 2001.

Last May, UNHCR announced that the group refugee status for Eritreans who fled their country as a result of the independence war or the recent border conflict with Ethiopia would end on 31 December 2002. It said the root causes of the Eritrean refugee problem no longer existed.

"Those Eritreans found to be still in need of international protection after undergoing individual screening will be able to remain in their current host country as refugees," UNHCR added.

"Those who do not qualify for asylum after 2002 but who do not wish to return home because of strong family, social or economic links with the host country will be expected to legalise their stay in Sudan, or the other countries where they currently reside," it said.

A further 5,000 Eritrean refugees are registered in Ethiopia, Yemen and Kenya.

ERITREAA–ETHIOPIA: FEATURE – ERITREAN DESERTERS IN "ENEMY" LAND

SHIRARO, ETHIOPIA, 22

Nov 2002 (IRIN) - In the dry, hilly landscape of northern Ethiopia, a group of young men wanders aimlessly through a makeshift "main street", sometimes stopping to take tea or to make conversation along the way.

IRIN; Eritrean refugees arrive in Shiraro.

They have nowhere to go and nothing to do. They are Eritreans - mostly deserters from the army or young people fleeing the military call-up at home. They have ended up in "enemy" territory, in a refugee camp for Eritreans.

The remote Wa'ala Nihibi camp, near the town of Shiraro, was originally set up to accommodate some 4,000 ethnic Kunamas from Eritrea who fled their country in 2000, at the height of the war with Ethiopia.

But this year, non-Kunamas started appearing. Small numbers of mostly young men began arriving at the camp after undergoing intensive screening by the Ethiopian authorities. As the months went by, the numbers started swelling - over 200 have crossed in the last two months alone, according to figures provided by the UN refugee agency, UNHCR, and the Ethiopian government.

Contingency Plans

The numbers are significant enough to be of concern to UNHCR which is drawing up emergency plans for a large influx of refugees from Eritrea.

UNHCR and the Ethiopian government's Administration for Refugee and Returnee Affairs (ARRA) say some 960 non-Kunamas have arrived so far this year - 823 of them Tigrinya speakers from the Eritrean highlands, and 65 of them women.

The majority are crossing near Adigrat in Ethiopia's northern Tigray region. In making the journey, the asylum seekers have to cross the

25 km buffer zone which separates the two countries following their bitter two-year border war.

"They are crossing because of forced military recruitment and persecution in Eritrea," says Berhe Woldemichael, the camp co-ordinator for ARRA, which offers protection to the refugees and administers the camp.

The mostly-uneducated Kunama and the draft dodgers - many of them students - are unlikely bedfellows in this barren camp, only a few kilometres from the Eritrean border and close to the controversial village of Badme where the war erupted in 1998. They rarely mix, living at opposite ends of the camp.

"Most of us are soldiers who have been taken from our homes, students forced to be soldiers," says a spokesman for the group. "We are grateful to the Ethiopian government and to ARRA, but there is nothing for us in this camp." They are bored, and he is afraid they will turn to alcohol because there is nothing to do.

"The Eritrean government is exploiting the brains and power of the youth," he alleges. "For them, we provide cheap labour. We are running in order to live in peace."

The deserters claim that the ages for national service in Eritrea - 18 to 40 - have either been lowered or increased. "Now, even if you are 60 you can be called up," the spokesman says. "People are being forced into service regardless of whether or not they have already completed 18 months of military service."

Eritrea Denies

The claims are vigorously denied by the Eritrean government, which says there is no second call-up.

"We are absolutely not changing the ages for national service," government sources told IRIN. "It's not at all sensible. We didn't do this at the height of fighting, so why would we do it now? Of course, every deserter will tell you such things."

The Eritrean government accuses the dominant party in Ethiopia's coalition government - the Tigray People's Liberation Front (TPLF) - of using the refugees for political gain.

"The figures given by the TPLF are nonsense," the source said. "The TPLF is just trying to make propaganda."

"People are dodging the draft for national service - it happens everywhere," he said. "This is a post-war situation - it's perfectly normal for a few draft dodgers to be there, and for their soldiers to be here."

"With this huge border, it's not surprising," he added. "Ethiopian soldiers are also crossing into Eritrea. We have their names. Eleven crossed two days ago, and nine some days before that."

Refugee Status Problematic

The refugees in Wa'ala Nihibi camp say they live in constant fear of attack by the Eritrean army. "There are security issues," admits Berhe, the camp administrator.

Wa'ala Nihibi is supposed to be a temporary camp, until a suitable site is identified further away from the border. A site was found, but the Ethiopian government said no to the move at the last minute. UNHCR, which is anxious to move the refugees, says another site is currently being evaluated.

Eritrea rejects suggestions that it is beefing up its military presence close to the border. "Here in Eritrea it's business as usual," the government source said.

The recent influx poses a problem for UNHCR which will end the blanket refugee status granted for Eritreans as of December 2002.

"There are several options," Ilunga Ngandu, the UNHCR representative in Ethiopia, told IRIN.

"Some may go home when relations between the two countries improve, but the conditions have to be right. Secondly, some may be settled locally in a viable environment. Or there could be resettlement in a third country for those who qualify."

It means that each person now crossing the border will have to be individually assessed before being granted refugee status and some could be sent home.

Eritrea says it will welcome

back its citizens. "If they return to Eritrea, they are coming home," the government said.

Eritrea: Reporters 'may have died in detention'

NAIROBI, 16 Nov 2006 (IRIN) - Three Eritrean reporters who have been in detention in a remote northeastern jail for five years are believed to have died in unclear circumstances, a global media freedom watchdog reported.

Reporters Without Borders, which has written to the Eritrean government seeking an explanation, said on Tuesday that the three were being at a place called Eiraeiro. "Dozens of political prisoners have disappeared into jails run by the armed forces," it added. "They include at least 13 journalists, of whom there has been no word for nearly five years."

Eritrean officials, who declined to comment on the report, have in the past denied allegations that the country holds political dissidents.

According to Reporters Without Boundaries, Eiraeiro, in the Sheib subzone of the Northern Red Sea administrative region, holds at least 62 political prisoners, including former ministers, senior officials, high-ranking military officers, government opponents and eight of the 13 journalists held since a round-up in September 2001. The journalists were captured by the police after the government decided to "suspend" all Eritrea's privately owned media.

Nine of the detainees at Eiraeiro, the watchdog added, had died as a result of "various illnesses, psychological pressure or suicide". Detainees in the prison are chained by their hands, sleep on the ground without bed linen, have their heads shaved once a month and are let out of their cells for an hour a day without being allowed contact with other prisoners, Reporters Without Borders said.

Eritrea–Ethiopia: Media watchdog deplores constraints on journalists

[This report does not necessarily reflect the views of the United Nations]

NAIROBI, 5 May 2004 (IRIN) - Fourteen journalists are being held behind bars in Eritrea, making the country the worst in Africa to work in as a journalist, an international media watchdog said on Monday.

The French-based Reporters Without Frontiers (RWF) said in its 2004 annual report that "little has changed" in the country despite pressure from the international community to improve conditions. "Nothing has shifted in Eritrea, still Africa's biggest prison for journalists and one of the last countries in the world without an independent, privately owned press," it said. "Pressure from the international community, including the European Union, proved ineffective."

According to the report, the government of Eritrea forced all privately owned newspapers to close, and imprisoned leading journalists in 2001. "Since then, the only source of news for Eritreans has been the government press and the few foreign radio stations that can be received," it said.

Yemane Gebremeskel, the director of the Eritrean president's office and presidential spokesman, told IRIN in an interview in April that "the problem with the existing papers was that they were few, most of the journalists were not experienced, they could have been easily manipulated, easily infiltrated, especially if there is money involved".

"If you tell me you are going to be a journalist, there are standards, there are ethics. In the previous press law, that was not there, so anybody who wanted to be a journalist could be a journalist. But then you also pay the price, because sometimes things get distorted," Yemane said in the interview.

The report also criticised neighbouring Ethiopia and named four journalists who were imprisoned in 2003. "The Ethiopian press still has to cope with great difficulties," it said. "Several journalists were arrested in 2003, and one was still being held at the end of the year. The government seems determined to adopt a new press law that will impose draconian restrictions on press freedom."

Ethiopian officials insist that the country has a flourishing private press and point to the independent newspapers operating there. The government spokesman, Zemedkun Tekle, told IRIN in March that the new press laws were debated openly and democratically.

RWF condemned the recent ban imposed on the Ethiopian Free Press Journalists Association (EFJA). "The EFJA's closure in November 2003 - ostensibly for purely bureaucratic reasons - indicated a new toughening in the attitude of the authorities," it said. "After allowing something of an opening in recent years, the government seemed to trying to reassert control and step up pressure on independent news media," it added.

Although the ban on the EFJA has now been lifted, bitter in-fighting between the former leadership and the new executive has plagued the organisation.

Sudan/Eritrea tensions put Eritrean repatriation on hold

NHCR News Stories; Unable to return home for now, Eritrean refugees in Showak camp, eastern Sudan, are growing into their host community. © UNHCR/R.Wilkinson.

KHARTOUM, Sudan, October 8 (UNHCR) Thousands of Eritrean refugees ISudan are in limbo as recent tensions between the two countries have put their repatriation on hold just months before their refugee status comes to an end. The UN refugee agency had originally planned to resume return convoys to Eritrea on Saturday, Oct 5, after a temporary break in June due to the rainy season. It had organised 16 trucks to restart the voluntary repatriation of Eritrean refugees living in and around the towns of Kassala and Showak in eastern Sudan.But last week, rebels believed to be from the Sudan People's Liberation Army (SPLA) attacked the towns of Homoshakarieb and Sholalab, north-east of Kassala. Sudan accused Eritrea of allowing the rebels to pass through their territory to launch the attack, an allegation the Eritrean government immediately rejected. The border was subsequently closed, and UNHCR was forced to suspend its plans to restart the return movement.The refugee agency was also asked to stop information campaigns currently underway in several camps in the area. Since Friday, Sudan has expelled 10 Eritrean government officials from the ministry that handles refugee/returnee matters. Five were asked to leave Gedaref state on Monday, while another five who were in Kassala state were expelled on Friday. The Eritrean government officials have been working with UNHCR on information campaigns in the camps and on registration of refugees who wish to return home.Over the weekend, restrictions were imposed on the movement of UNHCR staff in eastern Sudan, but they have now been lifted. The refugee agency has resumed registration for return in camps and in urban centres.The recent developments will set back UNHCR's plans to assist the voluntary return of more than 100,000 refugees still living in camps in eastern Sudan. It is the oldest large-scale refugee situation in the world with which the UN refugee agency is involved.In May this year, UNHCR declared the cessation of refugee status for Eritreans as of year's end. By then, many of those still living in Sudan will cease to be considered as refugees due to the fact that the original reasons for their flight – Eritrea's war of liberation and a subsequent conflict with Ethiopia – are no longer valid. Those wishing to remain in Sudan must apply for an alternative status or present any claims for continued asylum to a joint UNHCR/Government of Sudan panel for review.

Since the return movement started in May 2001, UNHCR has facilitated the repatriation of more than 50,000 camp-based refugees. However, more than 100,000 still remain in camps and several thousand more in urban centres. Story date: 8 October 2002. UNHCR News Stories

Eritrea: thousands more pour into Sudan

This is a summary of what was said by UNHCR spokesperson Kris Janowski – to whom quoted text may be attributed – at the press briefing, on 30 May 2000, at the Palais des Nations in Geneva.

Thousands of new Eritrean refugees Monday poured across the Sudanese border to Lafa, arriving on foot, donkey carts and trucks and tractors which shuttled them away from villages in western Eritrea that had reportedly been taken by Ethiopian troops.

UNHCR staff in Lafa this morning reported that the numbers crossing dwindled overnight, but that small groups were still arriving. They estimate that the total for the previous 24-hour period could reach 10,000.

Refugees arriving in Lafa told UNHCR that Ethiopian troops now controlled the town of Tesseney, 30 kms from the Sudanese border. Most said there had been no fighting for the town, although others reported artillery fire nearby. None of the arrivals Monday was injured. Refugees said that Ethiopian troops checked the identification papers of those who streamed out of the Tesseney area, apparently looking for soldiers, but otherwise did not stop or harass people taking to the road. Most of those crossing into Sudan Monday were children and women, although there were also significant numbers of men.

The latest arrivals are exhausted and badly in need of water and food. Temperatures in the arid border region soar past 40° Celsius during the day.

One 45-year old man died yesterday shortly after crossing at a point called Deman, near Umsafir.

UNHCR and partners are now operating 14 water

tankers between Kassala town and the border sites, and health posts are manned by MSF-Holland, the Islamic African Relief Association and the Sudanese Red Cross.

UNHCR dispatched another 700 tents and 400 jerrycans to the sites yesterday and is purchasing 5,000 jerrycans and more than 40,000 bars of soap from local suppliers. We will begin airlifting additional shelter material from our emergency stocks in Copenhagen tomorrow.

The fall of Tesseney was confirmed by a member of the Eritrean refugee commission who traveled there yesterday and returned to Sudan with his family. The towns of Talatashar, 13 kms inside the border, and El Kdir, east of Tesseney on the main road to Asmara, were also reported to be in Ethiopian hands.

Both Eritrean and Sudanese vehicles have been shuttling refugees across the border as word of the recent Ethiopian advance spread. On Sunday, 10 Sudanese trucks brought Eritreans from the towns of Gerghef (on the Eritrean side) and Gurji to Gerghef in Sudan. They are charging fees for the transfer.

The 600 UNHCR tents set up at Amara Musa, the site 2 kms from Lafa selected by local Sudanese authorities, are now all occupied. The site is urgently being expanded to accommodate the new arrivals, who are camping in the scrub just inside the border. In addition to a few dozen Ethiopians who have crossed into the Kassala area with the Eritrean refugees, UNHCR has registered a small number of Ethiopians who crossed overland to Port Sudan. We are working with Ethiopian officials to help them return to their country.

Story date: 30 May 2000.
UNHCR Briefing Notes

ERITREA–ETHIOPIA: FEATURE – PLIGHT OF THE KUNAMA REFUGEES

SHIRARO, 2 Dec 2002 (IRIN) - They call their homeland Kunama land – fertile plains that fall along the highly contentious border between Ethiopia and Eritrea. Yet many Eritrean

© IRIN Kuchi Raya and other Kunama refugees

Kunamas now find themselves living as refugees in a temporary camp in Ethiopia, only a few kilometres from the border.

"I love my country," says 57-year-old Kuchi Raya, a sentiment shared by dozens of the Kunama refugees taking shade from the sweltering heat under a large acacia tree. "It was not our intention to come here but what has happened in our country forced us to." Some 4,000 Kunama fled Eritrea in 2000 as the war with Ethiopia was drawing to a bloody close. Like many refugees they left all their possessions behind. Although the international community has largely ignored their plight, the complex and sensitive issues that surround these refugees are now coming to the fore.

Dilemma of the Camp

Crucial demarcation of the disputed 1,000-km border is due to take place early next year with expected territorial and population exchanges.

The dusty refugee camp of Wa'ala Nihibi, near the town of Shiraro, falls in one of the most hotly contested areas - Badme and the Yirga Triangle, the flashpoint of the two-year border war which erupted in 1998.

The camp's close proximity to the existing border with Eritrea also poses further dilemmas and has heightened tensions among the refugees.

"It makes us nervous," admits Kunama elder Afeworke Kalo, who fled after the withdrawal of the Ethiopian army from western Eritrea under the terms of the June 2000 ceasefire agreement. "We are worried about being so close. There is a suspicion of the Kunama by the Eritrean regime."

Puzzle of the Kunama Plight

There are estimated to be around 100,000 Kunamas in Ethiopia and Eritrea. They speak their own language and are visibly different from their highland neighbours. About 70,000 are said to live in Eritrea - mostly in the Gash Barka region.

Yet their plight is a puzzle. Tens of thousand remain in Eritrea. Those who fled are mostly the population of two villages whose districts fell under the control of the Ethiopian army.

Their flight, alongside the Ethiopian army which pulled back under the ceasefire agreement, sparked accusations that they had sided with the Ethiopians. Historically, Eritrea has questioned the Kunamas' support for independence from Ethiopia.

Kuchi, a mother of four, adds: "When the Ethiopians left, we were suspicious about revenge the Eritreans might take on us saying we helped the Ethiopians."

The Eritrean authorities say they are keen for the Kunama refugees to return home, arguing that they were taken to Ethiopia against their will.

"They left with the [Ethiopian] army," a government official told IRIN. "As far as we are concerned they are abductees. People don't go voluntarily with an invading army. And if they did go voluntarily, in the midst of war, then they are not refugees. They must be sent home."

He denied claims by the Kunama that they are "discriminated against" by the Eritrean government.

"The people of Eritrea, including most Kunamas, do not think they are being discriminated against," he said. "Those Kunama who are under the TPLF [Tigray People's Liberation Front, dominant party in Ethiopia's coalition government] are being forced to say such things."

The official denied claims by the Kunama refugees that the town of Barentu - the Kunama heartland in Eritrea - was made the capital of Gash Barka province as a "propaganda move" by the Eritrean government.

"Any law promulgated by

the government is not directed at one ethnic group only, it is for the benefit of all Eritreans," he stated. "By moving the provincial capital to Barentu [from Agordat], it was a clear sign of the interest the state has in that part of Eritrea and its people. The government does not take action for political propaganda."

Water carriers at Wa'ala Nihibi

Continued Refugee Status

The future of the Kunama refugees is further complicated by the impending December 2002 decision that brings to an end automatic refugee status for Eritreans around the world.

The UN's refugee agency, UNHCR, is considering exempting the Kunama at Wa'ala Nihibi from this – but in the meantime their temporary existence continues as they anxiously await their fate.

The UN's Mission in Ethiopia and Eritrea (UNMEE) is also monitoring the situation ahead of demarcation. They patrol in and around the 25 km buffer zone separating Ethiopia and Eritrea – an area that includes Wa'ala Nihibi. UNHCR is keen that the camp is shifted away from the border.

Both UNHCR and the Ethiopian government's Administration for Refugee and Returnee Affairs (ARRA) have tried to secure a new camp. One area was fitted out by UNHCR and ready for use last February, but the Ethiopian government vetoed the move at the last minute.

Another site is being looked at but the government's refugee administration – which has the final say – believes it will be at least another six months before a move can take place.

The new camp will require

infrastructure and facilities under laws that govern refugees. Food, shelter, healthcare and education will all need to be provided. As pastoralists living from cattle, the Kunama will also need grazing lands. UNHCR is also looking into the possibility of resettling the Kunama in a third country.

The fact that the camp is to be moved has also played a role in creating tensions. Some Kunama complain bitterly about inadequate relief assistance, a view shared by some officials working there.

Water carriers at Wa'ala Nihibi

But like many refugees they have gradually made the camp their home despite its obvious lack of facilities. Small shops selling a few vegetables and grain have sprung up.

Even so, currently there are 58 people to every toilet, although one non-governmental organisation, International Rescue Committee (IRC), supported by UNHCR, is building new ones.

IRC is also responsible for education in the camp and a small school has been constructed which provides teaching in several languages, including Kunama and Tigrinya.

Refugees Welcome to Go Home

Although the Kunama refugees admit they came to Ethiopia alongside its army, they are adamant they do not want to go home yet.

"I came here to save myself and my family," says Kuchi, who came from the village of Fode some 55 km away. Her view is echoed by fellow refugees. She complains that many of her children and relatives have gone missing after being "forced into military service".

"Until things change I do not want to return," she says. "What is it that would attract me to go back?"

The Eritrean government reiterates that it wants its citizens to go home. "If they come home, there is no way they will suffer recriminations," said the official.